3D Information Technologies for Tangible and Intangible Cultural Heritage

3D Information Technologies for Tangible and Intangible Cultural Heritage

Editors

Marek Milosz
Jacek Kesik

MDPI

Basel • Beijing • Wuhan • Barcelona • Belgrade • Novi Sad • Cluj • Manchester

Editors
Marek Milosz
Department of Computer Science
Lublin University of Technology
Lublin
Poland

Jacek Kesik
Department of Computer Science
Lublin University of Technology
Lublin
Poland

Editorial Office
MDPI AG
Grosspeteranlage 5
4052 Basel, Switzerland

This is a reprint of articles from the Special Issue published online in the open access journal *Applied Sciences* (ISSN 2076-3417) (available at: https://www.mdpi.com/journal/applsci/special_issues/information_technologies_cultural_heritage).

For citation purposes, cite each article independently as indicated on the article page online and as indicated below:

Lastname, A.A.; Lastname, B.B. Article Title. *Journal Name* **Year**, *Volume Number*, Page Range.

ISBN 978-3-7258-1613-2 (Hbk)
ISBN 978-3-7258-1614-9 (PDF)
doi.org/10.3390/books978-3-7258-1614-9

Cover image courtesy of Marek Milosz

Contents

applied sciences

MDPI

Editorial

Special Issue on 3D Information Technologies for Tangible and Intangible Cultural Heritage—Motivations, Overview of Published Works, and Perspectives for the Future

Marek Milosz * and Jacek Kęsik *

Department of Computer Science, Lublin University of Technology, ul Nadbystrzycka 36B, 20-618 Lublin, Poland
* Correspondence: m.milosz@pollub.pl (M.M.); j.kesik@pollub.pl (J.K.)

1. Motivation

The concept of tangible cultural heritage has been defined a long time ago and international legal acts have been created to protect, preserve, and popularize it. The most significant act is the "Convention Concerning the Protection of the World Cultural and Natural Heritage" adopted by the United Nations Educational, Scientific and Cultural Organization (UNESCO) in 1972 [1]. This convention obliges the members of the United Nations that have ratified it to protect their cultural and natural heritage. Material heritage is understood as a physical element of the heritage of humanity [1], and includes archaeological sites, historical monuments, artifacts, and other objects, as well as natural elements, such as physical and biological formations, geological and physiographical formations/areas, and natural sites/areas with high value for science, conservation, or natural beauty.

The intangible heritage of humanity was noticed much later. It was only on October 10, 2003 that the "Convention for the Safeguarding of the Intangible Cultural Heritage" was adopted by UNESCO [2]. Intangible cultural heritage is understood as, among others, "oral traditions, performing arts, social practices, rituals, festive events, knowledge and practices concerning nature and the universe or the knowledge and skills to produce traditional crafts" [3]. Intangible heritage understood in this way is inherited from our ancestors and should be passed on to our descendants as achievements of humanity. Intangible heritage is quite fragile because it mostly functions in human memory and if not "cultivated", it disappears, i.e., is forgotten. At the same time, it is a testimony to cultural diversity that should be maintained in the face of growing globalization and unification.

Older technologies, such as analogue photography, sound, and video recording, are now being replaced by digital ones. Information Technologies (ITs) bring a new quality to these activities. However, [4] points to the relatively low dissemination of IT applications in the area of the intangible heritage of humanity.

IT has been and can be used in various ways in the broadly understood field of cultural heritage. This use may include, but is not limited to, the following areas: documentation, protection, reconstruction, restoration, conservation, dissemination, and popularization.

Documenting is primarily the registration of the condition of an existing tangible artifact or of elements of intangible heritage. For this purpose, various IT tech can be used: examples include digital photogrammetry, 2D and 3D scanning using different technologies, and digital recordings of voice, image, and character movement [5–9].

Protection primarily entails the use of IT in the areas of physical protection of objects of tangible cultural heritage. For this purpose, all types of supervision and access protection systems are used [10,11].

Reconstruction is supporting the physical reconstruction of lost elements (e.g., vessels, paintings, or buildings) or conducting this kind of reconstruction in the digital realm (e.g., creating digital twins of currently non-existent cultural heritage artifacts) [12–14].

Citation: Milosz, M.; Kęsik, J. Special Issue on 3D Information Technologies for Tangible and Intangible Cultural Heritage—Motivations, Overview of Published Works, and Perspectives for the Future. *Appl. Sci.* **2024**, *14*, 4853. https://doi.org/10.3390/app14114853

Received: 29 May 2024
Accepted: 30 May 2024
Published: 4 June 2024

ITs can be used to support restoration and conservation works through spectral or statistical analysis of colours, materials, or technologies and through computer-aided planning of restoration/conservation activities [15,16].

Dissemination is the provision of information about cultural heritage to various social groups (e.g., scientists, guides, tourists). Modern ITs significantly reduce its costs and expand the range of recipients. They also increase the variety of data provided [17–19].

Popularization means providing a wide public with information about elements of cultural heritage. The IT tool used here is the Internet, thanks to various portals promoting cultural heritage [20,21].

In addition, the IT applications in the areas detailed above overlap with their uses in the field of scientific research, which concerns all areas: documentation, protection, reconstruction, restoration, conservation, dissemination, and popularization [22–24].

2. An Overview of Published Works

The following articles were published in the Special Issue titled "3D Information Technologies for Tangible and Intangible Cultural Heritage" (https://www.mdpi.com/journal/applsci/special_issues/information_technologies_cultural_heritage (accesed on 29 May 2024)):

A. Kęsik, J.; Miłosz, M.; Montusiewicz, J.; Israilova, N. Documenting Archaeological Petroglyph Sites with the Use of 3D Terrestrial Laser Scanners—A Case Study of Petroglyphs in Kyrgyzstan. *Appl. Sci.* **2022**, *12*, 10521. https://doi.org/10.3390/app122010521;
B. Montusiewicz, J.; Barszcz, M.; Korga, S. Preparation of 3D Models of Cultural Heritage Objects to Be Recognised by Touch by the Blind—Case Studies. *Appl. Sci.* **2022**, *12*, 11910. https://doi.org/10.3390/app122311910;
C. Kaplan, K. Real-Time Rendering Engines Help Visualize, Model, and Animate Ancient Cities: An Example in Antioch. *Appl. Sci.* **2022**, *12*, 12316. https://doi.org/10.3390/app122312316;
D. Kęsik, J.; Żyła, K.; Montusiewicz, J.; Miłosz, M.; Neamtu, C.; Juszczyk, M. A Methodical Approach to 3D Scanning of Heritage Objects Being under Continuous Display. *Appl. Sci.* **2023**, *13*, 441. https://doi.org/10.3390/app13010441;
E. Cruz Franco, P.; Rueda Márquez de la Plata, A.; Pérez Sendín, M. Investigating a Workflow for Obtaining Physical Models from Digital Twins Obtained through Photogrammetry and TLS: New Ways for a Sustainable Dissemination of Heritage. *Appl. Sci.* **2023**, *13*, 1057. https://doi.org/10.3390/app13021057;
F. Cao, M.; Jiang, H.; Zhao, H. Hash Indexing-Based Image Matching for 3D Reconstruction. *Appl. Sci.* **2023**, *13*, 4518. https://doi.org/10.3390/app13074518;
G. Way, D.; Wei, Y. Use of Cloud-Based Virtual Reality in Chinese Glove Puppetry to Preserve Intangible Cultural Heritage. *Appl. Sci.* **2023**, *13*, 5699. https://doi.org/10.3390/app13095699;
H. Kim, S.; Park, Y. 3D Reconstruction of Celadon from a 2D Image: Application to Path Tracing and VR. *Appl. Sci.* **2023**, *13*, 6848. https://doi.org/10.3390/app13116848;
I. Korga, S.; Dziedzic, K.; Skulimowski, S.; Gnapowski, S. Optimising Amber Processing Using 3D Scanning: New Perspectives in Cultural Heritage. *Appl. Sci.* **2023**, *13*, 12973. https://doi.org/10.3390/app132412973;
J. Diara, F.; Barsacchi, F.; de Martino, S. Moving beyond the Content: 3D Scanning and Post-Processing Analysis of the Cuneiform Tablets of the Turin Collection. *Appl. Sci.* **2024**, *14*, 4492. https://doi.org/10.3390/app14114492.

Among the published articles, only two out of ten (20%) discuss intangible cultural heritage, namely Chinese glove puppetry (article G) and the amber processing process (article I). Three articles (A, D, and J) focus on the documentation of objects of tangible cultural heritage, primarily using three-dimensional methods: terrestrial laser scanning and photogrammetry. Two articles are related to the reconstruction (articles F and H) of cultural heritage objects and two to their popularization (articles B and C). One article presents the use of IT for the dissemination of heritage, while making it accessible to visually impaired.

3. Conclusions and the Future

Among the articles included in this Special Issue regarding the use of 3D IT in the area of tangible cultural heritage, the crucial fields of protection, restoration, and conservation are notably under-represented. The current situation, with conflicts increasing all over the world, encourages us to undertake research in these areas. Analysing the trends among the most recent publications in the area of cultural heritage, we can hope that this gap will effectively be reduced in the coming years.

The gaps in the area of intangible heritage are even greater. In this Special Issue, there are no articles on uses of 3D IT in areas such as oral traditions and talks (including story-telling), performing arts (such as dancing, singing, celebrations, games, and plays), rituals, crafts, etc. These areas require advanced research and application work. Similarly to the case of tangible heritage outlined above, articles covering this area are starting to be present in the heritage science space, which suggests an increase in interest in these areas in the coming years, which we are looking forward to.

Regardless of fluctuations in the scientific community's areas of interest, new methods and 3D IT devices are being created or improved, including the following:

- mobile, air and UAV lidars;
- satellite imaging;
- geo-radars;
- motion capture systems.

Additionally, work is underway to improve the software for processing 3D data and spatial modelling obtained using the above methods. Meanwhile, artificial intelligence is entering various areas of science and technology, including everyday life, and will inevitably also reach the scope of heritage science.

This Special Issue of the journal *Applied Sciences* has been closed. However, the continuous development of research on and applications of 3D IT in the area of cultural heritage results in interesting new publications. The editors of *Applied Sciences* appreciate this branch of science and have decided to continue this scientific journey in an upcoming Special Issue titled "Challenges and Current Applications of 3D Information Technologies for Cultural Heritage" (https://www.mdpi.com/journal/applsci/special_issues/726R29M8E4 (accesed on 29 May 2024)).

We are pleased to invite all potential authors to consider it in their current or future academic endeavours.

Acknowledgments: We hereby thank all the authors and reviewers for their enormous and valuable contributions to this Special Issue.

Conflicts of Interest: The author declares no conflicts of interest.

References

1. Convention Concerning the Protection of the World Cultural and Natural Heritage. UNESCO 1972. Available online: https://whc.unesco.org/archive/convention-en.pdf (accessed on 27 May 2024).
2. Convention for the Safeguarding of the Intangible Cultural Heritage. UNESCO 2003. Available online: https://ich.unesco.org/en/convention (accessed on 27 May 2024).
3. What Is Intangible Cultural Heritage? UNESCO 2024. Available online: https://ich.unesco.org/en/what-is-intangible-heritage-00003 (accessed on 27 May 2024).
4. Skublewska-Paszkowska, M.; Milosz, M.; Powroźnik, P.; Łukasik, E. 3D technologies for intangible cultural heritage preservation—Literature review for selected databases. *Herit. Sci.* **2022**, *10*, 3. [CrossRef] [PubMed]
5. Tanasi, D.; Cardona, D.; Hassam, S.; Kingsland, K.; Kraft, M.; Trapani, P.; Calderone, D.; Giorgi, L.; Leucci, G. Remote sensing surveying at the Abbatija tad-Dejr Catacombs (Rabat, Malta). *Dig. Appl. Archaeol. Cult. Herit.* **2023**, *30*, e00271. [CrossRef]
6. Jebur, A.K. The Techniques of Cultural Heritage: Literature Review. *Saudi J. Civ. Eng.* **2022**, *6*, 108–114. [CrossRef]
7. Reshma, M.R.; Kannan, B.; Raj, V.J.; Shailesh, S. Cultural heritage preservation through dance digitization: A review. *Dig. Appl. Archaeol. Cult. Herit.* **2023**, *28*, e00257. [CrossRef]

8. Lim, V.; Khan, S.; Picinali, L. Towards a more accessible cultural heritage: Challenges and opportunities in contextualisation using 3d sound narratives. *Appl. Sci.* **2021**, *11*, 3336. [CrossRef]
9. Barszcz, M.; Montusiewicz, J.; Paśnikowska-Łukaszuk, M.; Salamacha, A. Comparative Analysis of Digital Models of Objects of Cultural Heritage Obtained by the "3D SLS" and "SfM" Methods. *Appl. Sci.* **2021**, *11*, 5321. [CrossRef]
10. Markiewicz, M. Image communication and contemporary visualisation in the popularisation of archaeology. *Spraw. Archeol.* **2022**, *74*, 2. [CrossRef]
11. Chen, F.; Zhou, W.; Tang, Y.; Li, R.; Lin, H.; Balz, T.; Luo, J.; Shi, P.; Zhu, M.; Fang, C. Remote sensing-based deformation monitoring of pagodas at the Bagan cultural heritage site, Myanmar. *Int. J. Digit. Earth* **2022**, *15*, 770–788. [CrossRef]
12. Cai, Z.; Fang, C.; Zhang, Q.; Chen, F. Joint development of cultural heritage protection and tourism: The case of Mount Lushan cultural landscape heritage site. *Herit. Sci.* **2021**, *9*, 86. [CrossRef]
13. Joshi, R.; Tamrakar, A.; Magaiya, B. Community-based participatory approach in cultural heritage reconstruction: A case study of Kasthamandap. *Prog. Disaster Sci.* **2021**, *10*, 100153. [CrossRef]
14. Montusiewicz, J.; Barszcz, M.; Dziedzic, K. Photorealistic 3D Digital Reconstruction of a Clay Pitcher. *Adv. Sci. Technol. Res. J.* **2019**, *13*, 255–263. [CrossRef]
15. Pietroni, E.; Ferdani, D. Virtual Restoration and Virtual Reconstruction in Cultural Heritage: Terminology, Methodologies, Visual Representation Techniques and Cognitive Models. *Information* **2021**, *12*, 167. [CrossRef]
16. Basu, A.; Paul, S.; Ghosh, S.; Das, A.; Chanda, B.; Bhagvati, C.; Snasel, V. Digital Restoration of Cultural Heritage With Data-Driven Computing: A Survey. *IEEE Access* **2023**, *11*, 53939–53977. [CrossRef]
17. Li, D.; Lin, H.; He, Q.; Li, J. Digital restoration and visualization of Nanfeng ancient city-focusing on cultural mining. *J. Phys. Conf. Ser.* **2021**, *1961*, 012045. [CrossRef]
18. Pei, J.; Yi, L. Fluid space: Digitisation of cultural heritage and its media dissemination. *Telemat. Inform. Rep.* **2022**, *8*, 100022. [CrossRef]
19. Khan, I.; Melro, A.; Amaro, A.C.; Oliveira, L. Role of Gamification in Cultural Heritage Dissemination: A Systematic Review. In *Proceedings of the Sixth International Congress on Information and Communication Technology*; Yang, X.S., Sherratt, S., Dey, N., Joshi, A., Eds.; Lecture Notes in Networks and Systems; Springer: Singapore, 2022; Volume 235. [CrossRef]
20. Cobo-Serrano, S.; Arquero-Avilés, R.; Marco-Cuenca, G. Accessibility and dissemination of the cultural heritage in a museum library: Case study in the National Archaeological Museum (Spain). *LIBER Q. J. Assoc. Eur. Res. Libr.* **2022**, *32*, 1–17. [CrossRef]
21. Cui, C.; Zhao, Y.; Wang, L. Protection and dissemination of Chinese intangible cultural heritage based on digital games. *Int. Commun. Chin. Cult.* **2021**, *8*, 483–491. [CrossRef]
22. De Luca, L. A digital ecosystem for the multidisciplinary study of Notre-Dame de Paris. *J. Cult. Herit.* **2024**, *65*, 206–209. [CrossRef]
23. Siatou, A.; Alexandrakis, G.; Pouli, P.; Curulli, A.; Kavoulaki, E.; Knezic, S.; Padeletti, G. A Multi- and Interdisciplinary Methodological Approach for Monitoring Cultural Heritage Built Assets: The HERACLES Experience. In *Advanced Nondestructive and Structural Techniques for Diagnosis, Redesign and Health Monitoring for the Preservation of Cultural Heritage*; Osman, A., Moropoulou, A., Eds.; Springer Proceedings in Materials; Springer: Cham, Switzerland, 2022; Volume 16. [CrossRef]
24. Kamariotou, V.; Kamariotou, M.; Kitsios, F. Strategic planning for virtual exhibitions and visitors' experience: A multidisciplinary approach for museums in the digital age. *Dig. Appl. Archaeol. Cult. Herit.* **2021**, *21*, e00183. [CrossRef]

applied sciences

MDPI

Article

Documenting Archaeological Petroglyph Sites with the Use of 3D Terrestrial Laser Scanners—A Case Study of Petroglyphs in Kyrgyzstan

Jacek Kęsik [1,*], Marek Miłosz [1], Jerzy Montusiewicz [1] and Nella Israilova [2]

1 Department of Computer Science, Lublin University of Technology, ul. Nadbystrzycka 36B, 20-618 Lublin, Poland
2 Computer Science Department, Kyrgyz State Technical University Named after I. Razzakov, Prospekt Mira, 66, Bishkek 720044, Kyrgyzstan
* Correspondence: j.kesik@pollub.pl

Abstract: The use of 3D terrestrial laser scanners (TLS) in the documentation of archaeological sites is an effective method of collecting information about the area under study. The wide range of acquired data makes this method a versatile tool, and not limited only to documentation tasks. This article presents the possibilities of 3D TLS and their postprocessing software in the pioneering work related to the digitization of exhibits in The Petroglyphs of Cholpon-Ata Open-air Museum near Lake Issyk-Kul in Kyrgyzstan. A 3DScaMITE methodology adapted for that task is highlighted. The data obtained during the scanning were used to build high-accuracy 3D digital petroglyph models, together with their location within the open-air museum area. The acquired models also allowed a detailed analysis of the geometric parameters of the cavities forming the petroglyph figures. The results of the analysis confirmed the thesis about improper preservation of petroglyphs. It has been demonstrated in this way that the used TLS method is completely universal in documenting petroglyphs, including the location and shape of their place of creation, as well as creating a sufficiently accurate analysis of the structure of drawings.

Keywords: 3D terrestrial laser scanning; 3D models; petroglyphs; geographic positioning visualization; analysis of petroglyph geometry

Citation: Kęsik, J.; Miłosz, M.; Montusiewicz, J.; Israilova, N. Documenting Archaeological Petroglyph Sites with the Use of 3D Terrestrial Laser Scanners—A Case Study of Petroglyphs in Kyrgyzstan. *Appl. Sci.* **2022**, *12*, 10521. https://doi.org/10.3390/app122010521

Academic Editor: David Mills

Received: 30 September 2022
Accepted: 16 October 2022
Published: 18 October 2022

1. Introduction

Petroglyphs, the rock drawings carved by representatives of different cultures throughout human history, are usually the work of people of the era before the creation of alphabets and inscriptions. They are a pre-literary form of presenting stories and ideas, in use from around the 10th millennium BC to almost modern times. Petroglyphs are found on all continents of the world except Antarctica, the continent considered uninhabited by humans. These drawings are perhaps the oldest evidence of human civilization and intelligence. Their exploration, documentation, and preservation for future generations is a very important element of protecting the cultural heritage (CH) of mankind.

Petroglyphs, as human rock art, occur mostly in currently little-known and not very frequented places, i.e., caves, mountain gorges, sea cliffs, rocks in desolate places, etc. Such places are usually not protected, and the nature of petroglyphs—rock drawings, does not allow them to be moved to safer places without irreversible damage to their surroundings. Petroglyphs are therefore exposed to various threats, leading to their loss for humanity, and also as a result of human activities.

The theft of rocks with petroglyphs occurs in practically all countries on all continents. For example, in California, USA, petroglyphs [1] made by the local Paiute tribe and held sacred by them, were stolen by cutting the stones they were carved on (Figure 1). A similar case took place in Nevada, USA [2] and Tasmania [3]. There are also reports of petroglyphs being treated as souvenirs by tourists [4]. In addition to stealing, acts of sheer vandalism against petroglyphs are noted. For example, vandals have irrevocably damaged petroglyphs in Arizona, USA [5] and Kyrgyzstan [6]. These are just examples of the destruction of petroglyphs, and there are more such cases all over the world.

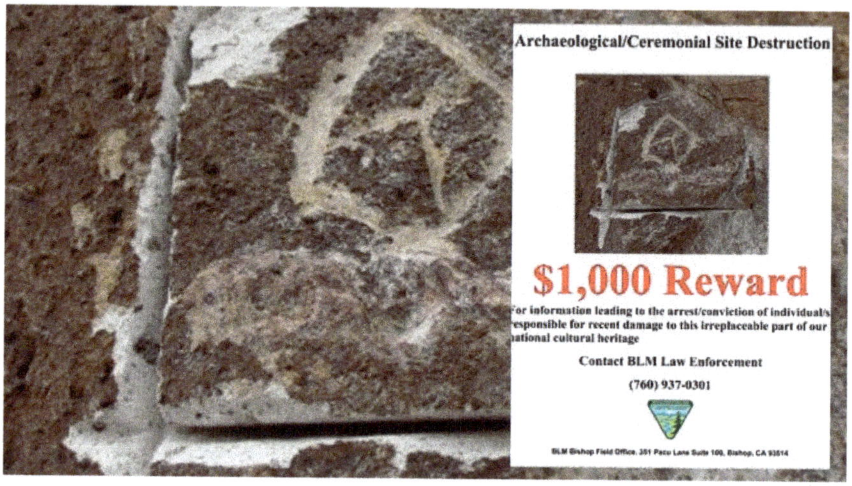

Figure 1. An example of a wanted poster for petroglyph theft [1].

Unfortunately, there are also errors of petroglyph conservators [7], leading to the partial destruction of monuments during restoration works. This leads to the need to remove incorrectly selected protective layers, which is not always possible, and always carries the risk of further damage to the monument. A description of the method of removing aged acrylic coatings (also used for the maintenance of petroglyphs) from wall paintings with the use of microemulsions, can be found in [8].

Such cases occur all over the world, and therefore urgent and permanent documentation of petroglyphs that have survived to this day should be a priority. This is made possible by the digital technology which consists of 3D scanning of their geometry, appearance and position. The ever more popular and affordable tool that could be used for this are 3D terrestrial laser scanners (TLS).

The purpose of this work includes:

I. Adaptation of the 3DScaMITE [9] methodology, developed for scanning cultural heritage objects with 3D TLS, to the scanning of petroglyphs in the places of their occurrenc
II. Verification of the proposed adaptation during the scanning of petroglyphs at The Petroglyphs of Cholpon-Ata Museum in Kyrgyzstan.
III. Verification of the usefulness of the obtained data to confirm or reject the thesis about improper conservation of petroglyphs.

The remaining part of the paper is organised as follows:

- Section 2 provides an overview of publications on documenting rock art and the methods of its protection;
- Section 3 discusses the main aspects of 3D TLS scanning, and presents modifications to the 3DScaMITE methodology for petroglyph scanning tasks;
- Section 4 discusses the effects of scanning petroglyphs at Cholpon-Ata and discusses the results in terms of confirming the thesis about the improper conservation of petroglyphs.

2. Rock Art Documentation

The issue of protection of various types of historical and cultural properties, apart from defensive ones, including sacred and secular structures and such objects as petroglyphs, inscriptions or stone monuments, has been taken up by state and local authorities in all corners of the world: Asia (Mongolia [10], Russia [11], United Arab Emirates [12]), Africa [13], North America [14], South America (Brazil [15], Peru [16]), Australia [17,18] and Europe (Spain [19–23], Sweden [24–26], Italy [27,28], UK [29] and France [30]). For example, according to the inventory conducted in Mongolia [10], petroglyphs and other rock art constitute 14% (i.e., 3261 artefacts) of the monuments in this country. Research on rock art (petroglyphs) is carried out primarily by domestic researchers working in "their backyard", although there are instances of highly advanced international collaborations [12], for example a three-year project led by Departamento de Ingeniería Topográfica y Cartografía of the Universidad Politécnica de Madrid (Spain) in the Emirate of Sharjah, United Arab Emirates.

The activities carried out currently take three main directions, apparently not always interconnected: cultural research, research on data acquisition methods, and protection and sharing. The basic direction of research is the analysis of petroglyphs in terms of culture, e.g., the assessment of zoomorphic and pragmamorphic artefacts in the context of the existence of certain rhetorical figures—assigning the properties of animals and objects to humans [24], or in the aspect of interpreting rock narratives in the context of the theory of migration and diffusion of ideas, along with using ethnographic and folklore materials [11]. The problem of dating rock sites has not yet been solved in numerical terms. It is easier to build a reliable chronological framework that takes into account aspects such as group mobility or emerging cultural networks [23].

In recent years, many studies have presented a description of various modern technologies for data acquisition in conducted research. In works [24,29] various technologies used to document petroglyphs are traced, from the oldest, often called traditional ones, which are now less and less used, such as frottage, tracing and drawing. All these technologies maintain fairly precise dimensions of the rock drawings, although they have many disadvantages: they poorly reflect spatial surfaces, are very time-consuming, are not suitable in cold and humid climates, and largely depend on the experience of the person carrying out the work. There is no reference to dimensions in the methods based on single photos obtained with conventional or digital photography, taken without placing additional markers or rulers on the objects. In the last two decades, many new digital technologies have been introduced in the acquisition of rock drawings: reflectance transformation imaging (RTI) [19,20], structure from motion (SfM) [24,31], photogrammetric measurements [21,22,28], optical laser scanning (OLS) [24,32], terrestrial laser scanning (TLS) [15,27,31], photogrammetric measurements combined with aerial recordings [12,16] using an unmanned aerial vehicle (UAV) or combining tachymeter-based tracking of the scanner, stereo and SfM [27]. RTI technology consists in taking approximately 60 photos of the object with a camera placed on a tripod, with changing light directions (evenly distributed on the hemisphere), which, after processing the collected data, leads to the collection of information about the value of the 2.5D model [19]. Other technologies, by using specialised software, allow for the more or less simple building of 3D models of petroglyphs, which significantly facilitates their storage and sharing. Article [22] is noteworthy, as it presents a detailed methodology describing all the stages of creating 3D petroglyph models using photogrammetry. Paper [14] is a methodological study on the advantages/disadvantages of using long-range laser scanning and high-resolution digital photography for petroglyph recording, and in [18] the method of combining spectral information from cave wall scanning by means of geodetic measurements is presented, with petroglyphs from digital elevation mode (DEM). Having a very large collection of rock art allows one to conduct research aimed at using artificial intelligence by building a neural network for the automation of identification and classification of rock art images [26].

The issue of protecting petroglyphs and making them available to local and international communities appears more and more widely in the scientific literature. In [13] the authors pay attention to the fact of adapting the used petroglyph digitisation methods to the economic potential of the country, while in [22] low-cost solutions for the preparation of 3D models of petroglyphs are presented. The need to create online platforms to integrate geo-referenced information with generated 3D petroglyph models for the purpose of more effective analysis and data availability, which facilitates further scientific research, can be found in [12,28]. Aspects of the usefulness of digital technologies (photogrammetry and TLS) for creating 3D models to protect rock art objects against their loss due to the deterioration of their condition, because of the variability of weather conditions and possible disasters (natural—earthquakes, international—military actions, or social—theft, devastation), can be found in [18,30,33]. Attention is also paid to the social responsibility of professional communities dealing with 3D digitisation of petroglyphs and other cultural monuments for promoting these techniques, so that their beauty can be recognised and examined by others in the future [18]. Paper [17] analyses the advantages and disadvantages of laser scanning, photogrammetry and photographic reconstruction in capturing and managing knowledge of rock art sites, with particular emphasis on practical applications in the field of cultural heritage management, with the example of West Angelas, in the East Pilbara region of Western Australia.

3. Contemporary 3D TLS and the Methodology of Their Use for Documenting Archaeological Artefacts

When performing 3D scanning and documentation of archaeological artefacts, two aspects that have a significant impact on obtaining the intended results can be distinguished: the selection of the appropriate equipment, and the use of an appropriate scanning methodology. In the case of scanning petroglyphs, the use of laser technologies allows for the scanning of the artefact with its natural surroundings, during one scanning session consisting of many (3–5) scans.

3.1. TLS Scanners

Contemporary 3D terrestrial laser scanners are universal and fully automatic measuring devices. They differ in their purpose and range of scanning. The general principle of operation of these devices is based on the measurement of the distance from the device to the object, on the basis of the analysis of the reflection of the laser beam from the surface of the object. Most devices use a near infrared laser. The maximum TLS scanning range can be divided into short-range—up to 70/150 m (e.g., Surphaser 100HSX, FARO Focus S70), medium-range—up to 350 m (e.g., Trimble TX8, FARO Focus S350, Leica RTC360), and long-range, where the range is measured in kilometres (e.g., Teledyne Optech Polaris, Riegl VZ-6000).

The basic data obtained is a cloud of 3D points of the places where the laser reflects from the objects around the scanner, recorded in the coordinate system relative to the position of the scanner. In addition to the coordinates of the reflection point, the scanner can measure the energy of the reflected signal, which allows for the determination of the brightness level of the surface at any given point (Figure 2a).

Laser detection alone does not allow the assigning of a colour to the detected surfaces. Most of the modern TLS devices also have the ability to take a spherical photograph of the environment (Figure 2b), on the basis of which the scan is coloured. There are three groups of solutions:

1. Use of an external camera/digital camera mounted in a special holder and synchronized with the scanner.
2. Use of a built-in spherical camera or set of cameras permanently installed in the scanner housing.
3. Use of a built-in camera using the same optical path as the laser measurement.

Figure 2. Exemplary data obtained with TLS scanning (excavations in Lublin, Poland): (**a**) a three-dimensional point cloud with a brightness level, (**b**) panoramic image of the surroundings to colour the scan. Source: authors' own.

The advantage of the first method is the possibility of using a popular (reasonably priced) digital camera with parameters matching the required effect; the disadvantage is the need to calibrate the camera after changing the parameters (e.g., focal length). Solutions two and three eliminate the need for calibration, as the cameras are already factory calibrated. The disadvantage is the rigid limitation of the resolution of the obtained spherical image, resulting from the physical parameters of the built-in devices. In the case of solutions one and two, additionally, errors in the assignment of colours may occur, due to the difference in the position of the laser emission point and the focal point of the spherical image (Figure 3). These errors are most significant in close proximity to the device. Solution three eliminates this problem, due to the positioning of the camera in the axis of the laser beam emission.

The obtained coloured scan (3D point cloud with colours) is saved in coordinates relative to the position of the scanning device. In order to be able to determine the location of the scanned areas in the real world, it is necessary to obtain data on the geographical location of the scanner, along with its height above sea level, the rotation angle of the scanner's base position in relation to the north direction, and the direction and angle of the scanner from the horizontal position. With this information, it is possible to pre-position and align multiple scans of a given area.

Depending on the model of the device, TLSs are equipped with various sensors allowing the obtaining of the above-mentioned data. Determining the geographical position of the scanner is performed with the use of the global navigation satellite system (GNSS) sensor module. Currently, the GPS, BeiDou, and Galileo networks are available worldwide. Depending on the module used by the TLS, it can pinpoint its position using one or more networks, which results in an average outdoor positioning accuracy ranging from 2 to 10 m. Actual accuracy can be diminished by signal interference.

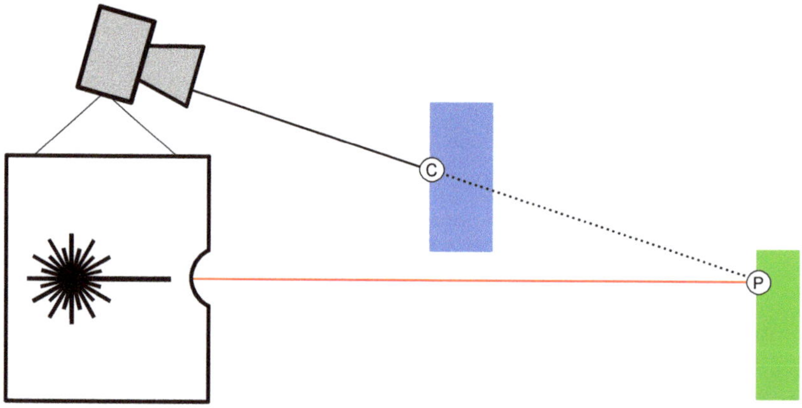

Figure 3. Occurrence of an error in assigning a colour to a specific position of an object's surface. Designated surface position—P, colour reading location—C.

Depending on the solution used in the device, the altitude above sea level is obtained on the basis of readings from GNSS data or by using an electronic barometer built with MEMS (MicroElectroMechanical Systems) technology. The MEMS technology also creates a magnetometer, used by the scanner as an electronic compass. The compass reading is used to pre-align the obtained scan with respect to the north direction. In order to measure the inclination (inclinometer), a MEMS accelerometer is used, which measures the changes in induction caused by the inclination of the mass in the sensor in response to changes in the position of the device in relation to the gravitational field of the earth. The tilt reading is used to level the scan.

Modern TLS scanners are therefore able, during a single measurement session, to obtain clouds of object surface measurement points, data on their levelling, height above sea level, GPS position and orientation towards the sides of the world (Figure 4), as well as photographic images of high resolution and colour quality of the scanned surfaces.

Performing a series of such automated measurements in the field allows for quick and highly accurate acquisition of geometric and visual data that allow archaeological artefacts to be documented.

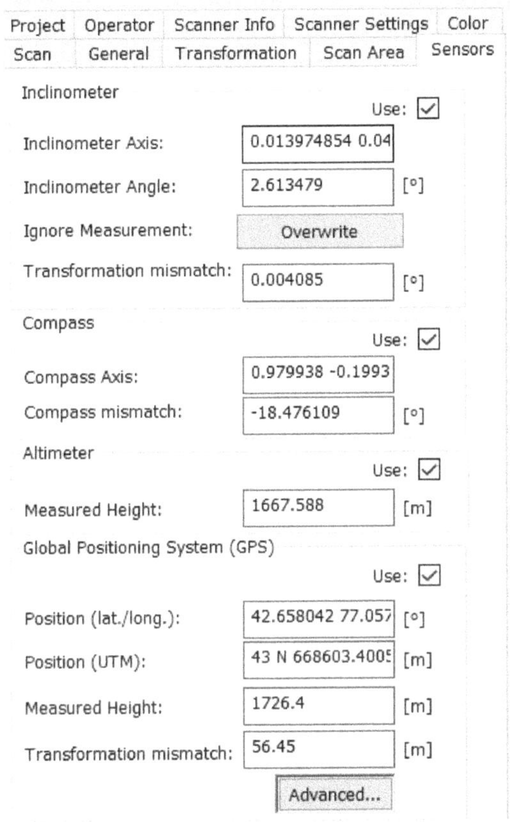

Figure 4. Sensor panel of the FARO Focus TLS.

3.2. Scanning Methodology

When carrying out the task of scanning cultural heritage objects with the use of TLS scanners, it is necessary to follow a specific methodology. This methodology should enable the scanning work to be carried out in a way that, after processing the obtained data, would not force repetition of the actions. The 3DScaMITE methodology of scanning cultural heritage objects with TLS scanners presented by the authors in [9], distinguishes the following stages: planning, implementation, results processing and dissemination of results. The methodology modified for petroglyph scanning tasks is presented in Table 1.

The suitability of the scanner used in the research (Faro Focus X330) has been assessed based on scanner specifications. Although the scanner has a systematic error of 1 mm at 20 m (repeatable for each scanner unit), this refers to the global position of the petroglyph surface. The error of measuring petroglyph carvings depth is represented by the random measurement error (noise), which is 0.3–0.5 mm, depending on the surface reflection. This error is additionally reduced by the scanner hardwired procedure of repeating the measurement of each point multiple times (increasing the beam exposure time for each point). This, compared with the expected petroglyph carvings depth of 1–8 mm and the scanner-to-surface distance within the 1.5–5 m range, allows for the claim that the scanner used is suitable for distinguishing petroglyph carvings.

Table 1. Adapted 3DScaMITE [9] methodology for scanning petroglyphs.

	Name of the Stage	Sub-Stages	Results
1.	Planning	1.1. Strategic planning 1.1.1. Getting to know the petroglyph site 1.1.2. Selecting petroglyphs to scan 1.1.3. Acquiring photos of the petroglyphs and surroundings 1.1.4. Determining the order of petroglyph scanning	Sequence and location of petroglyphs to be scanned—initial list.
		1.2. Initial planning 1.2.1. Determining number and positions of scans per petroglyph chosen. Setting scanning sequence 1.2.2. Setting scanner parameters 1.2.3. Setting naming-conventions data	Final list of objects. Initial plan of scanning of every petroglyph chosen.
		1.3. On-site replanning 1.3.1. Identifying expected problems 1.3.2. Rearranging scan plan 1.3.3. Marking problematic positions	Plan of scanning adjusted to on-site conditions.
2.	Scanning	For each petroglyph from list: For each scan position: 2.1 Place and set parameters 2.2. Perform scan 2.3. Check scan quality 2.4. Mark problematic scan (if any, go to 2.2) 2.5. Acquire additional data	Folders for each petroglyph containing: files with scans data, additional data (photo, historical, …).
3.	Data processing	3.1. Preliminary processing 3.1.1. Process chosen scans, initial rough scan placement based on sensors read 3.1.2. Check data quality 3.1.3. Identify scans suitable as donors (if needed)	Source point clouds arranged over site area.
		3.2. Full processing 3.2.1. Filter and clean up data 3.2.2. Remove or substitute problematic areas 3.2.3. Align partial scans based on sensors and cloud to cloud compare 3.2.4. Deduplicate points	Final point clouds of each petroglyph arranged over site area.
		3.3. Final processing For each petroglyph: 3.3.1. Set processing region 3.3.2. Perform conversion 3.3.3. Choose corresponding supplementary data	Base (high quality) model of each petroglyph with corresponding supplementary data (e.g., background images).
4.	Preparing for dissemination	4.1. Obtain requirements for dissemination model 4.2. Convert base to dissemination model 4.3. Test quality (go to 4.2 if needed) 4.4. Arrange dissemination petroglyph model in viewing environment, using supplementary data. 4.5. Prepare/finalise model for distribution	Dissemination of petroglyph model.

4. Petroglyphs at Cholpon-Ata, Kyrgyzstan and Their Scanning

One of the most famous stone fields with rock drawings is located above the shore of Lake Issyk-Kul (approx. 1630 m above sea level) in the village of Cholpon-Ata in Kyrgyzstan. The stone field was created by a break in a mountain lake embankment several thousand years ago, which allowed the lake waters to flood the valley. Hence, there are various types of large boulders and round logs lying there, unparalleled in the local landscape. Located there is an open-air museum called the Issyk-Kul Provincial State Historical Cultural Museum-Reserve. The museum covers a stone field with an area of

42 ha with thousands of stones with drawings dating from as early as the 2nd millennium BC to the 8th century AD [34]. The images were mainly made by the nomadic Turkic tribes called Saka and Usun. They contain zoomorphic themes: "depictions of goats and deer with ornamented bodies, camels, dogs, deer, boars, bulls, horses, wild rams, or predators: wolves; hunting scenes, juxtaposed animals, individual and paired goats, multi-figure compositions with goats in rows, geometrised goats, camel caravans, individual hunting scenes" [34]. A view of the petroglyph field open to visitors in the museum is shown in Figure 5.

Figure 5. A stone field with petroglyphs in the museum near a village of Cholpon-Ata. Source: authors' own.

Unfortunately, Kyrgyzstan has very limited resources to protect its cultural heritage. Thanks to projects sponsored from abroad, limited works related to the protection and reconstruction of monuments are carried out. This causes, among others things, a lack of permanent protection of the Issyk-Kul Provincial State Historical Cultural Museum-Reserve, and a high risk of its vandalism or destruction.

As part of the 3rd Scientific Expedition of the Lublin University of Technology to Central Asia, which took place in 2019 [35], a pilot scan of selected objects with petroglyphs in the museum was carried, out using the TLS scanner (Figure 6).

Unfortunately, the museum's poor protection against weather conditions (high mountain climate) and tourists, causes the petroglyphs to gradually deteriorate. As mentioned by some authors, e.g., [36], a significant number of stones in petroglyphs have been lost over the last fifty years, due also to mundane activities such as building roads and houses.

In addition, in 2002 over 300 stones were restored, including 74 with the most interesting drawings. In the course of this work, without preliminary tests, the rock drawings were covered with an acrylic resin called Paraloid B72 [7]. At first it seemed that the conservation had paid off—the drawings became more visible. But in a couple of years, under the influence of the high-mountain climate (high levels of UV radiation, large temperature changes, wind, etc.), the situation deteriorated significantly. The paraloid layer became less transparent, and in many places the surface of the stones began to peel off and then fall off [7]. The depressions forming the drawings originally had clear and even edges, and a depth of 1–3 mm (and sometimes more—up to 8 mm). After conservation, they were flooded and levelled with resin. The edges have become blurry and the depth of the carvings has decreased.

Figure 6. Implementation of the scanning of petroglyphs in the museum in Cholpon-Ata. Source: authors' own.

Paraloid B72 has been used in many places around the world to protect monuments. Currently, there is more and more talk about its harmfulness and the need to remove aged coatings [37–39].

5. Scanning Results and Discussion

Following the indicated modified methodology (Table 1), selected petroglyphs were scanned at the Issyk-Kul Provincial State Historical Cultural Museum-Reserve in Cholpon-Ata in Kyrgyzstan. The scanning procedure was carried out using the Faro Focus X330 scanner. From among the exhibits, it was decided to scan the five objects presented in Figure 7.

Work on the 3D digitization of petroglyphs has been carried out at the Issyk-Kul Provincial State Historical Cultural Museum-Reserve for the first time in history. It made it possible to document the appearance and condition of the museum's most valuable exhibits. The results (base data, digital models and their parameters), will be transferred to the museum's authorities for eternal storage and use in further scientific research.

The number and location of the scanning positions for each object were determined by the team of researchers, taking into account the size of the scanned object and the location of the images of the drawings—petroglyphs. Table 2 shows the GPS positions of these places obtained with the scanning device. Since all the petroglyphs are located in the Issyk-Kul Provincial State Historical Cultural Museum-Reserve, their location is already widely known; thus, presenting the GPS coordinates does not increase the risk of theft.

Figure 7. Petroglyphs selected for scanning. Presented numbers of petroglyphs are used in the following text. Source: authors' own.

Table 2. GPS scanner positions while obtaining petroglyph data.

Object Number	Latitude	Longitude
1 (4 positions)	42.658102	77.057090
	42.658042	77.057133
	42.658067	77.057237
	42.658157	77.057150
2 (2 positions)	42.658253	77.057732
	42.658222	77.057780
3 (3 positions)	42.658823	77.056525
	42.658845	77.056553
	42.658873	77.056488
4 (3 positions)	42.659255	77.056722
	42.659273	77.056762
	42.659302	77.056722
5 (3 positions)	42.659370	77.057098
	42.659387	77.057155
	42.659435	77.057122

The visual arrangement of the objects and the position of the scanner on the premises of the museum are presented in Figure 8. The location and positions for scanning the objects were determined on the basis of the GPS scanner position readings.

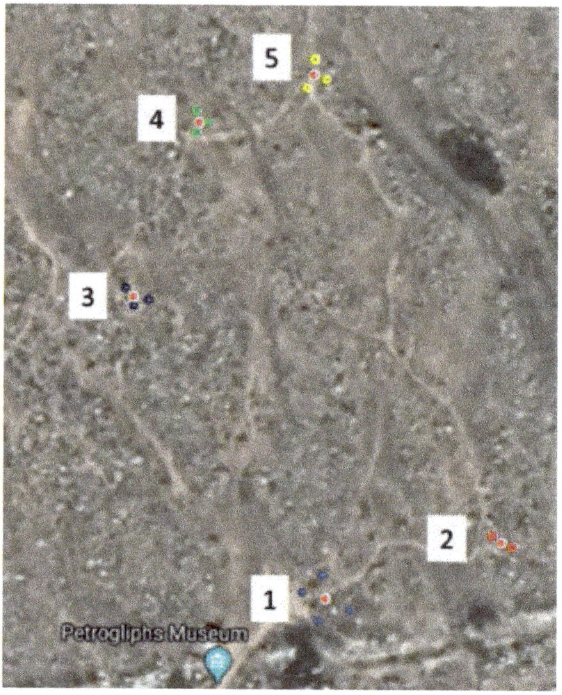

Figure 8. Arrangement of petroglyphs and assigned scan positions (coloured points around the positions of objects—marked with red dots with white halo). Source: own study with the use of Google Maps.

As a result of the scanning, clouds of points for individual boulders with petroglyphs were obtained, and then base 3D meshes were generated on this basis. Figure 9a shows an example of the obtained point cloud with the visualisation of the point density. The final result in the form of a generated 3D mesh, together with the visualisation of the mesh density, is shown in Figure 9b.

Table 3 gathers information on the total amount of vertices in the point cloud and the average point cloud density on the petroglyph surface for each object, together with the number of images of the generated mesh and the size of the base file. The density of the acquired point cloud makes it possible to detect carvings forming petroglyphs.

The 3D meshes are base ones which can be a source for dissemination models that, together with other data (e.g., GPS coordinates), can be used for different types of presentations, including virtual reality [40,41].

Table 4 presents a summary of the surfaces of individual boulders containing petroglyphs. The colour of the surface as well as the presence of carvings and unevenness are shown. It can be observed that the indentations corresponding to the petroglyph drawings are visible only for objects 1 and 5. In the case of object 2, a faintly distinguishable flat dimple can be observed in the area of the drawing (indicated by a red line). All objects show irregularities corresponding to the roughness of the rock surface.

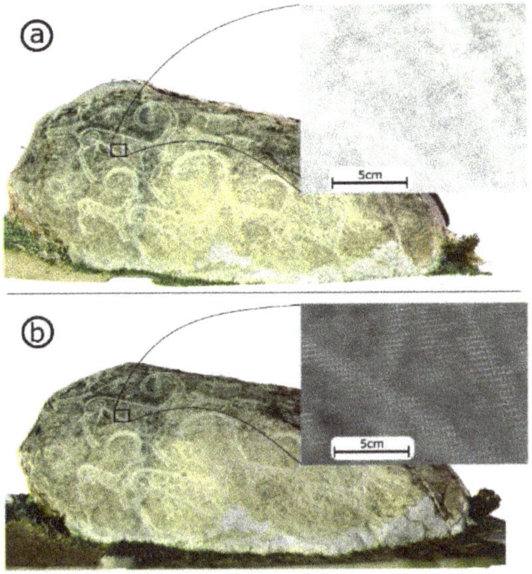

Figure 9. Examples of scanning results: (**a**) point cloud and (**b**) triangle mesh.

Table 3. Parameters of point clouds and generated 3D meshes.

Object No	Vertices in Point Cloud Count (M)	Average Point Cloud Density (pts/cm^2)	Mesh Faces Count (M)	File Size (MB)
1	5.534	107	2.547	132
2	5.015	114	2.823	149
3	3.794	118	2.431	130
4	3.441	120	3.473	185
5	4.457	140	5.085	274

In order to determine the depths of the pits in the models corresponding to the petroglyphs, cross-sections through the point clouds were made in the places where pits were found. The pit depth was calculated by triangulating the measurements of the distance between the bottom and pit edge points (Figure 10).

Information on the occurrence and depth of cavities in individual objects is collected in Table 5.

Figure 10. Visualisation of the carving depth determination method.

Table 4. The visibility of carvings in petroglyph drawings.

Object No	Fragment with Texture	Surface of the Fragment
1	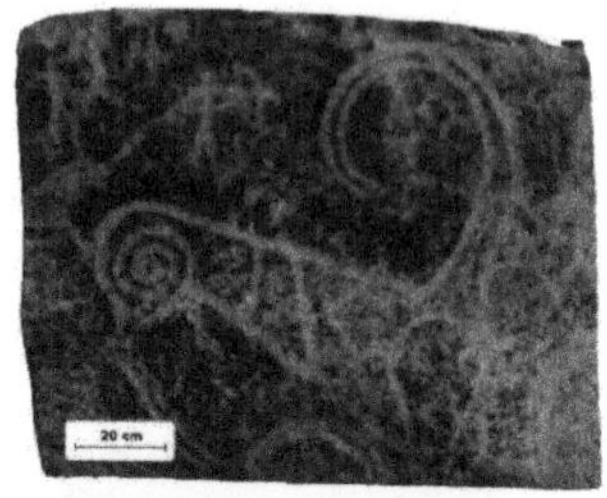	
2	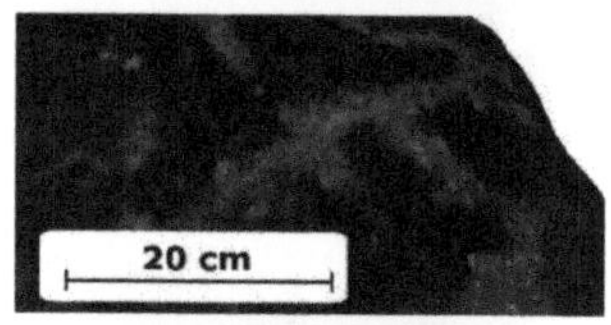	
3	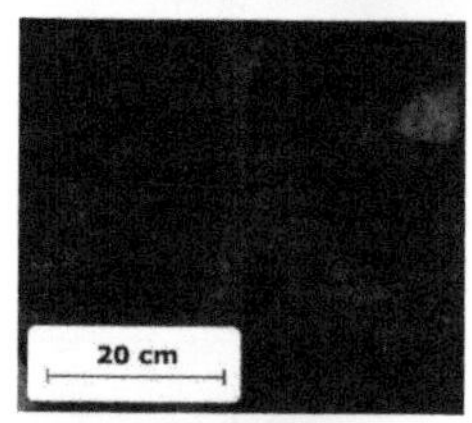	
4	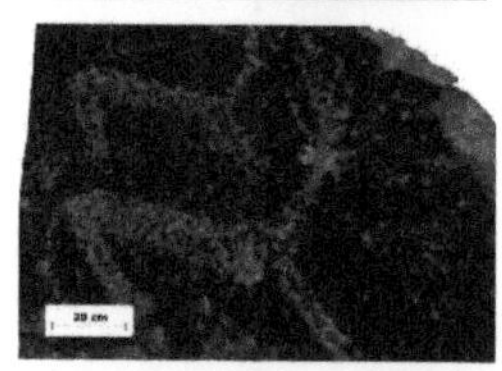	
5		

Table 5. Petroglyph carvings observed on the scanned object surface.

Object No	Carvings-Visibility	Carvings-Occurrence	Carvings-Average Depth (mm)
1	Clearly visible	Significant	1.5
2	Barely seen	Slight	0.4
3	Imperceptible	None	-
4	Imperceptible	None	-
5	Clearly visible	Significant	1.3

The obtained results confirm the thesis about improper conservation of petroglyphs. The original depressions of the rock drawings were levelled to such a significant extent that, in the case of some petroglyphs, they do not geometrically stand out on of the structure of the rock surface. Where depressions can be observed, they are significantly shallower. The observation of the edges of the indentations is difficult and in many cases impossible without the image of the colour of the rock surface.

At the same time, the scans reveal the roughness of the rock surface beyond the areas of the drawings, with details much smaller than the original size of the grooves forming the petroglyphs, which dismisses the suspicion that the lack of visibility of the grooves is caused by too low resolution and accuracy of scanning.

6. Conclusions

In the light of the conducted research, the following conclusions can be drawn:

Terrestrial laser scanner technology enables accurate documentation of petroglyphs, including their 3D geometry, external appearance, GPS position, and location in relation to the north/south direction, which allows for precise positioning of these objects on satellite photos and/or digital maps.

The possibilities of modern 3D TLS and the developed methodology for their effective use are presented in the example of the performed three-dimensional scanning of petroglyphs at the Issyk-Kul Provincial State Historical Cultural Museum-Reserve in Cholpon-Ata in Kyrgyzstan.

The obtained results of scanning five petroglyphs, i.e., the lack of indentations practically visible in mesh models (petroglyphs no. 3 and 4) resulting from carving the drawing on the surface of the stone, confirms the reports that the petroglyphs were poorly restored. The thesis that the coatings of the maintenance layers became less transparent as a result of aging, and the contours of the recesses were blurred, was confirmed

Unfortunately, the museum authorities were unable to provide information about the used method of petroglyphs preservation (e.g., method of film application, layer thickness, number of layers etc). This prevents the possibility of comparing the acquired results with other existing reports on the preservation of petroglyphs in other museums and their results.

Before future maintenance activities which should clean petroglyphs from the wrong protective layer, an accurate inventory should be carried out in order to obtain complete information, in the digital version of 3D models using modern technologies for 3D digitization.

During the research, a small number of the Issyk-Kul Provincial State Historical Cultural Museum-Reserve petroglyphs were scanned. The obtained data made it possible to verify the correctness of the developed scanning methodology and the high quality of its results. The next steps should be to scan the remaining artifacts in order to obtain a complete digital documentation of the petroglyph field. Digital models could be used by scientists from different disciplines for further research (e.g., to show different tools or techniques used for creating the rock art, to detect erosion patterns, etc.) without the need to study petroglyphs on site, for example [42].

Author Contributions: Conceptualization, M.M.; Data curation, N.I.; Formal analysis, M.M.; Investigation, N.I.; Project administration, J.M.; Resources, J.M. and N.I.; Software, J.K.; Visualization, J.K.; Writing—original draft, J.K.; Writing—review & editing, M.M., J.M. and N.I. All authors have read and agreed to the published version of the manuscript.

Funding: The work was co-financed by the Lublin University of Technology Scientific Fund FD-20/IT-3/007.

Institutional Review Board Statement: Not applicable.

Informed Consent Statement: Not applicable.

Acknowledgments: The authors thank the authorities of the Lublin University of Technology in the persons of former Rector: Piotr Kacejko, former Dean of the Faculty of Electrical Engineering and Computer Science: Danuta Stryczewska, and Head of the Department of Computer Science: Dariusz Czerwiński, for co-financing scientific expeditions of the Lublin University of Technology to Central Asia.

Conflicts of Interest: The authors declare no conflict of interest.

References

1. The "Wanted" Poster for Petroglyph Theft, Released in Fox News Site. Available online: https://static.foxnews.com/static/managed/img/Scitech/damage%20to%20rock%20art%20Sierra%20Nevadas.jpg (accessed on 15 October 2022).
2. Thevenot, C.G. Stolen Petroglyph Returns to Canyon after Rocky Journey. Las Vegas Journal. 2011. Available online: https://www.reviewjournal.com/crime/courts/stolen-petroglyph-returns-to-canyon-after-rocky-journey/ (accessed on 15 July 2022).
3. Ogilvie, F. The Art of Stealing. ABS News. 2019. Available online: https://www.abc.net.au/news/2019-11-10/aboriginal-petroAvailableonline:glyphs-stolen-and-missing-from-tasmania/11663382?nw=0 (accessed on 15 July 2022).
4. Vasyukova, Y. Petroglify Khabarovskogo Kraya Rastaskivayut na Suveniry a Politsiya Bessil'na. HabInfo. 2017. Available online: https://habinfo.ru/petroglify-razvorovyvayut/ (accessed on 15 July 2022).
5. Ryan, S. Vandalism and Petroglyphs. Archaeology Southwest. 2020. Available online: https://www.archaeologysouthwest.org/2020/03/16/vandalism-and-petroglyphs/ (accessed on 17 July 2022).
6. Vandalizm nad Petroglifami v Oshe—Informatsiyu Napravyat v Militsiyu. Available online: https://ru.sputnik.kg/society/20200915/1049625110/kyrgyzstan-petroglif-vandalizm.html (accessed on 20 March 2022).
7. Aleksandrov, Y. Khoteli kak luchshe ... Kirgizskiye "Restavratory" Sterli Unikal'nyye Petroglify. CentrAsia 2004. Available online: https://centrasia.org/newsA.php?st=1083311700 (accessed on 15 October 2022).
8. Taube, M.; Chelazzi, D. The removal of aged acrylic coatings from wall paintings using microemulsions. In Proceedings of the 17th Triennial Conference 2014 Melbourne Murals Stone and Rock Art 2014, Melbourne, Australia, 15–19 September 2014.
9. Milosz, M.; Kęsik, J.; Montusiewicz, J. 3D Scanning and Visualization of Large Monuments of Timurid Architecture in Central Asia—A Methodical Approach. *J. Comput. Cult. Herit.* **2021**, *14*, 31. [CrossRef]
10. Enkhbat, G. The Creation of a Registration and Information Database for Cultural Heritage in Mongolia. In *New Horizons for Asian Museums and Museology*; Sonoda, N., Ed.; Springer: Singapore, 2016. [CrossRef]
11. Леванова, Е.С.; Король, Г.Г. Основные Направления Научных Изысканий Е.Г. Дэвлет. *Краткие Сообщения Института Археологии* **2019**, *257*, 408–421.
12. Angás, J.; Jasim, S.A.; Bea, M.; Farjas, M.; Uribe, P.; Lanau, P.; Otero, X. Rock Art Recording in Khatm Al Melaha (United Arab Emirates): Multirange Data Scanning and Web Mapping Technologies. *Int. Arch. Photogramm. Remote Sens. Spat. Inf. Sci.* **2019**, *42*, 85–92. [CrossRef]
13. Tomášková, S. Digital technologies in context: Prehistoric engravings in the Northern Cape South Africa. *Digit. Appl. Archaeol. Cult. Herit.* **2015**, *2*, 2–3. [CrossRef]
14. Hurst, S.; Louden, E.; Johnson, E. Unravelingthe Cowhead Mesa petroglyphs with 3-D laser scanning and high-resolution digital photography. *Rock Art.* **2009**, *26*, 45–53. [CrossRef]
15. Kipnis, R.; Santos, H.B.C.; Tizuka, M.M.; Almeida, M.J.G.T.; Corga, M.P. Application of 3D modeling technologies combined with traditional techniques for registering rock engravings in the Madeira River, Rondônia, Brazil. *Bol. Mus. Para. Emílio Goeldi. Ciências Hum.* **2013**, *8*, 605–619. [CrossRef]
16. Berquista, S.; Spence-Morrow, G.; Gonzalez-Macqueen, F.; Rizzuto, B.; Álvarez, W.Y.; Bautista, S.; Jennings, J. A new aerial photogrammetric survey method for recording inaccessible rock art. *Digit. Appl. Archaeol. Cult. Herit.* **2018**, *8*, 46–56. [CrossRef]
17. Davis, A.; Belton, D.; Helmholz, P.; Bourke, P.; McDonald, J. Pilbara rock art: Laser scanning, photogrammetry and 3D photographic reconstruction as heritage management tools. *Herit. Sci.* **2017**, *5*, 25. [CrossRef]
18. Fryer, J.G.; Chandler, J.H.; El-Hakim, S.F. Recording and Modelling an Aboriginal Cave Painting: With or Without Laser Scanning? in TS5—3D Techniques for Heritage Applications, ISPRS Archives. 2018. Available online: https://www.isprs.org/proceedings/XXXVI/5-W17/pdf/3.pdf (accessed on 10 September 2022).
19. Díaz-Guardamino, M.; Sanjuán, L.G.; Wheatley, D.; Zamora, V.R. RTI and the study of engraved rock art: A re-examination of the Iberian south-western stelae of Setefilla and Almadén de la Plata 2 (Seville, Spain). *Digit. Appl. Archaeol. Cult. Herit.* **2015**, *2*, 41–54. [CrossRef]
20. Díaz-Guardamino, M.; Wheatley, D. Rock art and digital technologies: The application of Reflectance Transformation Imaging (RTI) and 3D laser scanning to the study of Late Bronze Age Iberian stelae. *Rev. Prehist. Andal. J. Andal. Prehistory* **2013**, *4*, 187–203.
21. Bendichoa, V.M.L.-M.; Ortegab, Á.M.; Vincentc, M.L.; Martín-Buitragoa, Á.J.C.; Pintadoa, J.O. Uso Combinado de la Fotografía Digital Nocturna y de la Fotogrametría en Los Procesos de Documentación de Petroglifos: El Caso de Alcázar de San Juan (Ciudad Real, España). *Virtual Archaeol. Rev.* **2017**, *8*, 64–74. [CrossRef]
22. Sanz, J.O.; Docampo, M.D.L.L.G.; Rodríguez, S.M.; Sanmartín, M.T.R.; Cameselle, G.M. A Simple Methodology for Recording Petroglyphs Using Low-Cost Digital Image Correlation. *J. Archaeol. Sci.* **2010**, *37*, 3158–3169. [CrossRef]

23. Ochoa, B.; García-Diez, M.; Domingo, I.; Martins, A. Dating Iberian prehistoric rock art: Methods, sampling, data, limits and interpretations. *Quat. Int.* **2021**, *572*, 88–105. [CrossRef]
24. Horn, C. The cunning of the fox: A case of zoomorphism in Scandinavian rock art. In *Giving the Past a Future: Essays in Archaeology and Rock Art Studies in Honour of Dr Phil. h.c. Gherhard Milstreu*; Dodd, J., Meijer, E., Eds.; Archaeopress: Oxford, UK, 2018; pp. 143–152.
25. Horn, C.; Ling, J.; Bertilsson, U.; Potter, R. By All Means Necessary—2.5D and 3D Recording of Surfaces in the Study of Southern Scandinavian Rock Art. *Open Archaeol.* **2017**, *4*, 81–96. [CrossRef]
26. Horn, C.; Ivarsson, O.; Lindhé, C.; Potter, R.; Green, A.; Ling, J. Artificial Intelligence, 3D Documentation, and Rock Art—Approaching and Reflecting on the Automation of Identification and Classification of Rock Art Images. *J. Archaeol. Method Theory* **2021**, *29*, 188–213. [CrossRef]
27. Höll, T.; Holler, G.; Pinz, A. A Novel High Accuracy 3D Scanning Device for Rock-Art Sites. *Int. Arch. Photogramm. Remote Sens. Spat. Inf. Sci.* **2014**, *40*, 285. [CrossRef]
28. Forgia, V.; Sineo, L. Within the symbolic world of the Paleolithic hunters: A GIS-based and 3D model analysis of sites with complexes of linear incisions in western Sicily. *Digit. Appl. Archaeol. Cult. Herit.* **2021**, *20*, e00175. [CrossRef]
29. Trinks, I.; Díaz-Andreu, M.; Hobbs, R.; Sharpe, K.E. Digital rock art recording: Visualising petroglyphs using 3D laser scanner data. *Rock Art Res.* **2005**, *22*, 131–139.
30. Lesvignes, E.; Robert, E.; Valentin, B. Using digital techniques to document prehistoric rock art: First approaches on the engraved panels of the Paris Basin shelters. *Digit. Appl. Archaeol. Cult. Herit.* **2019**, *15*, e00122. [CrossRef]
31. Zainuddin, K.; Majid, Z.; Ariff, M.F.M.; Idris, K.M.; Abbas, M.A.; Darwin, N. 3D Modeling for Rock Art Documentation using Lightweight Multispectral Camera. *Int. Arch. Photogramm. Remote Sens. Spat. Inf. Sci.* **2019**, *42*, 787–793. [CrossRef]
32. Horna, C.; Pitmanb, D.; Potterc, R. An evaluation of the visualisation and interpretive potential of applying GIS data processing techniques to 3D rock art data. *J. Archaeol. Sci. Rep.* **2019**, *27*, 101971. [CrossRef]
33. Fuentes-Portoa, A.; García-Ávilab, C.; Marrero-Salasc, E. Casa del samarín, una estación de grabados rupestres en deterioro. documentación, análisis y diagnóstico en los llanos de ifara, Granadilla, Tenerife. *Virtual Archaeol. Rev.* **2020**, *12*, 99–114. [CrossRef]
34. Jarosz, K. The Attractiveness of Cholpon Ata Resort and the Museum of Petroglyphs as a Tourist Destination. *Folia Tur.* **2017**, *44*, 9–29. [CrossRef]
35. 3rd Scientific Expedition of the Lublin University of Technology to Central Asia. Part 1—Kyrgystan. 13–17 May 2019. Available online: https://cs.pollub.pl/3-naukowa-ekspedycja-politechniki-lubelskiej-do-centralnej-azji-czesc-1-kirgistan-13-17-maja-2019/?lang=en (accessed on 20 September 2022).
36. Hermann, L. *Les Pétroglyphes de Tcholpon-Ata au Kirghizstan*; BoD-Books on Demand France: Paris, France, 2010.
37. Baglionia, M.; Benavidesa, Y.J.; Desprat-Drapelab, A.; Giorgia, R. Amphiphile-based nanofludis for the removal of styrene/acrylate coatings: Cleaning of stucco decoration in the Uaxactun archeological site (Guatemala). *J. Cult. Herit.* **2015**, *16*, 862–868. [CrossRef]
38. Giné, A.B.; Ortega, S.M.; Pons, P.R. The removal of Paraloid® coatings with aqueous based formulations. Practical case in frescoes from els Munts Roman villa (Catalonia). *Estud. Conserv. Restauro* **2019**, *10*, 67–83.
39. Baglioni, M.; Giorgi, R.; Berti, D.; Baglioni, P. Smart cleaning of cultural heritage: A new challenge for soft nanoscience. *Nanoscale* **2012**, *4*, 42–53. [CrossRef]
40. Salwierz, A.; Szymczyk, T. Methods of creating realistic spaces—3D scanning and 3D modelling. *J. Comput. Sci. Inst.* **2020**, *14*, 101–108. [CrossRef]
41. Ciekanowska, A.; Kiszczak-Gliński, A.; Dziedzic, K. Comparative analysis of Unity and Unreal Engine efficiency in creating virtual exhibitions of 3D scanned models. *J. Comput. Sci. Inst.* **2021**, *20*, 247–253. [CrossRef]
42. Wojcicki, P.; Korga, S.; Milosz, M. Preliminary Application of the Algorithm Highlighting Petroglyph Patterns. *Appl. Sci.* **2022**, *12*, 1660. [CrossRef]

applied sciences

MDPI

Article

Preparation of 3D Models of Cultural Heritage Objects to Be Recognised by Touch by the Blind—Case Studies

Jerzy Montusiewicz, Marcin Barszcz * and Sylwester Korga

Department of Computer Science, Faculty of Electrical Engineering and Computer Science, Lublin University of Technology, Nadbystrzycka 36B, 20-618 Lublin, Poland
* Correspondence: m.barszcz@pollub.pl

Abstract: Providing access to and the protection of cultural goods—intangible and tangible heritage—is carried out primarily by institutions such as museums, galleries or local cultural centres where temporary exhibitions are shown. The international community also attempts to protect architectural objects or entire urban layouts, raising their status by inscribing them on the UNESCO World Heritage List. Contemporary museums, however, are not properly prepared to make museum exhibits available to the blind and visually impaired, which is confirmed by both the literature studies on the subject and the occasional solutions that are put in place. The development of various computer graphics technologies allows for the digitisation of cultural heritage objects by 3D scanning. Such a record, after processing, can be used to create virtual museums accessible via computer networks, as well as to make copies of objects by 3D printing. This article presents an example of the use of scanning, modelling and 3D printing to prepare prototypes of copies of museum objects from the Silk Road area, dedicated to blind people and to be recognised by touch. The surface of an object has information about it written in Braille before the copy-making process is initiated. The results of the pilot studies carried out on a group of people with simulated visual impairment and on a person who is blind from birth indicate that 3D models printed on 3D replicators with the fused filament fabrication technology are useful for sharing cultural heritage objects. The models are light—thanks to which they can be freely manipulated, as well as having the appropriate smoothness—which enables the recognition of decorative details present on them, as well as reading texts in Braille. Integrating a copy of an exhibit with a description about it in Braille into one 3D object is an innovative solution that should contribute to a better access to cultural goods for the blind.

Keywords: 3D models; blind people; cultural heritage; 3D printing; recognition by touch; Braille alphabet

Citation: Montusiewicz, J.; Barszcz, M.; Korga, S. Preparation of 3D Models of Cultural Heritage Objects to Be Recognised by Touch by the Blind—Case Studies. *Appl. Sci.* **2022**, *12*, 11910. https://doi.org/10.3390/app122311910

Academic Editor: Asterios Bakolas

Received: 17 October 2022
Accepted: 18 November 2022
Published: 22 November 2022

1. Introduction

Museums across the world are institutions that primarily collect various types of material heritage collections. Some of them are made available in the form of permanent exhibitions, thematic collections or temporary exhibitions. Most people perceive art and museum exhibitions in a visual context: an exhibit, an explanatory text of the perceived object, more recently an individual receiver with a recording in the national language of the visitor or a QR code that allows one to connect via the Internet to the actual recording on one's own smartphone. Traditional exhibitions are widely available for healthy people and people with mobility impairments. However, not everyone is able to receive such displays. The real challenge is to make at least some of the exhibits accessible to visually impaired people. The classic way of securing exhibits items at exhibitions is by placing them in tightly closed cabinets (the easiest way to protect them against damage or theft), which means that visually impaired or blind people have no chance of getting acquainted with them.

It is currently estimated that there are one billion people worldwide with some form of blindness or visual impairment [1], and the number is increasing. Eye diseases cause limitations in everyday life as well as exclusion from social and cultural life. The vast majority of museums are not adapted to serve people with visual impairments. There is a lack of technical preparation, proper labelling of museum rooms, qualified personnel, proper form of information about events and, most importantly, the possibility of a kinaesthetic knowledge of objects.

According to European Union (EU) law, public institutions are required to adapt their institutions to welcome disabled people. This applies to both people with physical disabilities and those with sight and hearing impairments. However, these regulations are not respected, which does not mobilise directors of institutions to introduce changes. The problem is not the financial issues, but the insufficient flow of information about the available technologies for creating displays for the blind [2]. A description of the various activities of museums in the EU and around the world that try to adapt their displays for blind and visually impaired people is presented in Chapter 2. Adapting museums for visually impaired people is relatively inexpensive. It does not require the construction of special driveways, the installation of elevators or any major architectural changes.

For many people, a visit to a museum can make them feel left out. The traditional display of museum items in glass cases behind glass, despite the information written in Braille, is completely insufficient for the blind or visually impaired person (light reflections significantly impede the observation of the object). The authors encountered such examples in places inscribed on the UNESCO World Heritage List, e.g., in the Shahrisabz Museum of History and Material Culture (Shahrisabz, Uzbekistan), or a museum created in the area of Greco-Roman excavations in Nea Paphos (Cyprus). A much better solution for exposing exhibits to the blind can be found in the museum of Registan (Samarkand, Uzbekistan). The object is a fragment of the famous mosaics on the front of the Sher-Dor Madrasah ivan showing lions, gazelles and the sun, as seen in Figure 1a. The exhibit for the blind is a miniature whose surface has a different texture depending on which object it represents. This solution allows one to feel the outline and shape of individual objects. The facility for the blind is also enriched with an additional description in Braille placed next to the exhibit, as seen in Figure 1b.

(a)

(b)

Figure 1. Mosaic from Sher-Dor Madrasah, Registan Ensemble (Samarkand): (**a**) real object; (**b**) miniature for tactile cognition intended for visually impaired people.

Each museum has a wealth of fascinating objects that can be presented to viewers in a new convention. Audio and Braille descriptions are good branding tools for the exhibits on display and can help visitors better understand the purpose and history of an object. Contemporary exhibitions must also take care of activating other senses, for example, touch or smell, which will further enhance the experiences of visitors with a visual impairment. Personal contact with a museum artefact—kinaesthetic cognition or the narration of a guide who is able to paint vivid mental pictures—is much more attractive to a person. Museums must communicate with the visually impaired. They need to know their problems and listen to and work with them to understand what their real needs are. Thanks to these changes, museums can make the blind and partially sighted feel socially needed and make them a part of society [3].

The aim of this article is to develop a methodology for activities using 3D computer technologies to prepare copies of historic objects intended to be known by touch for people with visual impairments. This work presents copies made using 3D printing in FFF (fused filament fabrication) technology with the use of 3D modelling and the 3D scanning process. An important aspect of this article is that descriptions made in Braille were incorporated onto the surfaces of the 3D model copy, and furthermore, a pilot study was carried out with the participation of a person who was blind from birth.

2. Related Works

Some museums meet these problems by adapting their exhibitions to the blind and visually impaired and try to introduce to their offer various new solutions to make their exhibits available. The authors of [4] propose a multidisciplinary approach of supplementing the information about historic buildings on museum websites with content for visually impaired people. Such action, however, does not allow for a direct contact with the described object. Research conducted with the participation of visually impaired people shows that museum websites have a low degree of functionality [5]. It is possible for these people to get to know historic objects only by touching them [6], but most museums do not allow such a solution due to the risk of damage to the exhibit.

Study [7] reports research on the behaviour of visitors interacting with replicas of historic museum objects made with additive technology. The research has so far been conducted without the participation of blind people at the Burke Museum of Natural History and Culture in Seattle. For the purposes of the research, copies of four small museum objects were prepared, the size of which did not exceed 15 cm. The obtained results confirm that getting to know the exhibits using the sense of touch is an encouraging and interesting activity, and the copies made with the use of additive technology are safe and work well in practice.

In many museums around the world, low light intensity is used, which is intended to create an appropriate atmosphere of the exhibition, known as the play of lights. Viewing exhibits in such conditions is not convenient for many people, especially if one additionally needs to read information about the object. This situation is burdensome for elderly people with severe myopia. The management of many modern museums are aware about this problem and try to solve it in many different ways. The Mary Rose Museum, located in Historic Dockyards in Portsmouth, UK, hosts monthly dedicated shows for the visually impaired. On this day, the level of light is higher than usual, the sound effects are adapted to the environment and specially trained staff help visitors. Moreover, the museum also uses tactile materials, offering copies of its exhibits as 3D printed models [8]. The Tiflologico Museum located in Madrid, Spain, is a tactile museum created by the Spanish National Organisation for the Blind (ONCE—Organización Nacional de Ciegos de España). The main goal of this museum is to offer blind or partially sighted people normal access to the exhibits by creating a so-called museum without a visual barrier. The museum offers a wide variety of exhibitions, including models of famous buildings and the history of Braille and tactile art (created by visually impaired artists) that can be viewed both with the sense of sight and kinaesthetic cognition [9]. Similar exhibitions are organised by the Victoria and Albert Museum (V&A)—London's largest arts and crafts museum—which was founded in 1852 as the South Kensington Museum [10]. Once a month, the museum offers a descriptive presentation and a tour of its car exhibition especially for visually impaired people. The museum also offers extensive information (also available online), dedicated guide services for the blind, Braille books and audio descriptions as well as a multisensory backpacks containing objects to be known by touch, such as 3D models [11]. Direct contact with the exhibits is also possible in the DeCordova Sculpture Park (Lincoln, NE, USA), where many contemporary outdoor sculptures are located. Visually impaired people can visit the Sculpture Park through a touch tour [12]. The National Gallery in Prague (Czechia) offers visually impaired visitors the opportunity to "touch" in a virtual world three sculptural masterpieces considered as the largest in the world. Thanks to the appropriate feedback gloves, blind and visually impaired people can "touch" in the VR world to discover digital replicas of: Nefertiti's heads, Venus de Milo and Michelangelo's David [13]. Specialised educational programmes for the blind are implemented in Turkey. "Istanbul Modern The Colour I Touch" is a programme for blind and partially sighted children and adolescents. It allows for guided tours of exhibitions, participation in specialist workshops and screenings of films with audio descriptions [14]. The Van Loon House Museum (Amsterdam, The Netherlands) focuses on the kinaesthetic cognition of the exhibits. The museum is open 7 days a week and offers guided tours where one can touch the selected exhibits from the shared collection [15]. The Touch Museum, established in 1984 and located in Kallithea (Greece), in the historic building "Lighthouse for the Blind of Greece" has collected many interesting museum artefacts that can be explored kinaesthetically. All the objects in the museum are copies of the originals exhibited in other museums around the country and, importantly, they can be explored/experienced by touch [16]. The City Museum Trier ("Stadtmuseum Trier") [17] introduced audio tracks to tactile reproductions and replicas of museum exhibits dedicated to the visually impaired and blind. Article [18] describes the construction of a device intended for visually impaired people. It consists of a ring put on a finger, NFC sensors placed on the surface of the printed 3D model (in FFF technology) and an application for a tablet or smartphone. During kinaesthetic cognition, coupling occurs between the ring and the sensor, which activates the soundtrack through an application running on the smartphone. The authors used scanning and 3D printing to prepare the 3D model.

Study [19] is theoretical and cognitive in nature and concerns collecting opinions from blind people visiting museums in order to collect guidelines for creating tactile exhibits. The authors pay special attention to obtaining the appropriate copy surfaces of the exhibits and the materials used to make them. The authors of [20] emphasise the importance of 3D printing technology in museology, which in the future will lead to a paradigm shift in the museum model. In the future, museum objects or their copies will be accessible by touch. Multisensory forms of communing with culture will be important for the accessibility of cultural heritage, especially for people with learning difficulties, children, the elderly, blind or partially sighted people. As for today, it should be said that the standards for the transfer of 3D objects to multi-sensory feeling have not yet been developed. The proposed solutions require refinement, testing and examination. Publication [21] describes an interactive audio guide using a set of cameras identifying the depression of the analysed surface. The device shown was used to analyse the surface of copies of historic reliefs printed with FFF (fused filament fabrication) technology. The interaction between the finger depressions penetrating the relief and the distance read from the camera provides 2.5D spatial information, which results in the transmission of information in the form of audio. The constructed prototype was pre-tested in pilot studies with the participation of visually impaired people. The process of creating 3D models of museum objects using the Structure from Motion (SfM) method is presented in [22]. The authors presented the limitations of the automatic digitisation process (correction of the 3D model mesh, textures, etc.) and the method of verifying the compliance of the resulting digital copy of the exhibit with the original. However, the topic of 3D printing of replicas of museum artefacts was not discussed in the work. In [23], the authors used 3D scanning to generate copies of real digital models for kinaesthetic cognition through 3D printing (FFF technology). The digital models were post-processed to increase the recesses and emphasise the protruding elements. Printed models of sculptures and an architectural object were tested with a blind person. A prototype of an interesting solution is presented in [24]. The 3D haptic model of the historic Wenceslas Hill in Olomouc (Czechia) was made with 3D printing technology. In addition, the interaction between the user and the object was achieved by applying layers of conductive material responding to touch on selected surfaces of the model. Activating connections by touch leads to the display of detailed information on the tablet located next to the 3D model.

Creating replicas of museum objects, even with complex shapes, using 3D printing technology is possible thanks to the use of 3D scanning and 3D modelling. Technical and practical aspects concerning the use of various scanning techniques for historic museum exhibits were discussed in previous works [25–30]. Three-dimensional scanning using structured light technology was used to scan the statue of Hercules from the Antalya Museum in Turkey and the Khmer head from the Rietberg Museum in Zurich [25]. This work also presents the issues of selecting 3D scanning techniques for historic museum exhibits in terms of costs, experience and training of staff. Article [26] concerns checking the suitability of stationary and mobile 3D laser scanners for transferring archaeological objects to the digital world. The research confirmed the inability to collect information about the texture, and in the case of a mobile scanner there was a need to stick markers, and in the case of a stationary scanner there were limitations regarding the size of the artefacts. The authors of [27], apart from the aforementioned aspects, also deal with the issues of digital reconstruction. Work [28] presents a low-cost small object 3D data acquisition system consisting of a set of Kinect v2 devices cooperating with the open source LiveScan3D software. The system is difficult to transport over long distances (for example, by air transport) due to its large number of components. In article [29], the authors describe a significant number of available technologies and devices that in various situations can be used to acquire 3D data of large and small objects of tangible cultural heritage. Monograph [30] presents both various theoretical aspects concerning the issues of 3D digitisation of architectural objects and museum artefacts, as well as many practical implementations. Another approach, alternative to 3D scanning technology, is described by the authors of [31]. The article discusses the creation of digital architectural objects through the use of classic 3D modelling. The replicas made with the 3D printing technology were used to better understand the kinaesthetic in the board game.

The authors of the articles discussed used 3D models printed with FFF technology for kinaesthetic research, but in the conducted survey research in the form of pilot studies, they did not have a group of blind people. The analysis of the articles shows that the issue of making tangible cultural heritage objects available to blind people is a problem that has not yet been properly resolved, both in terms of organisation and technology.

3. Materials and Methods

3.1. The Use of 3D Printing Technology in Museology

People with visual loss increasingly expect to be able to experience visiting a museum in a similar way to a person with normal eyesight. Scientific research shows that the most important senses receiving stimuli from the environment are the senses of sight and touch [32]. Many museums around the world were closed between 2020 and 2021 due to the COVID-19 pandemic [33]. Due to this situation, the availability of museum exhibitions for the blind was drastically limited. Some of the museums made available selected objects of their resources by posting photos on their websites. This method is ineffective. However, there is another option to make the collections available. The availability of 3D printing technology can significantly contribute to the fact that blind people are also fully fledged people visiting museums or experiencing copies of museum artefacts, in their own homes or local cultural centres. Preparing to download files of 3D models of museum exhibits in the STL format makes it possible to print them for oneself. Thus, it is possible to create a private collection of the most famous and sublime works of art. Some museums provide these files for free. Thanks to the idea of museology as an "open source museum", the material cultural heritage is consciously shared.

3.2. The Process of Making Museum Exhibits Available

The procedures for making museum objects available are quite complicated, which is why they are presented in the form of a diagram in Figure 2. The diagram shows the developed procedure that allows for the presentation of museum objects in various forms: digital and real copies. The proposed procedure uses 3D scanning technology to obtain digital data about objects and transform them into digital 3D models along with the texturing process. For several years, the authors have been using the 3D scanning technology of museum objects both to protect cultural property and cultural heritage, and to make it available [34,35].

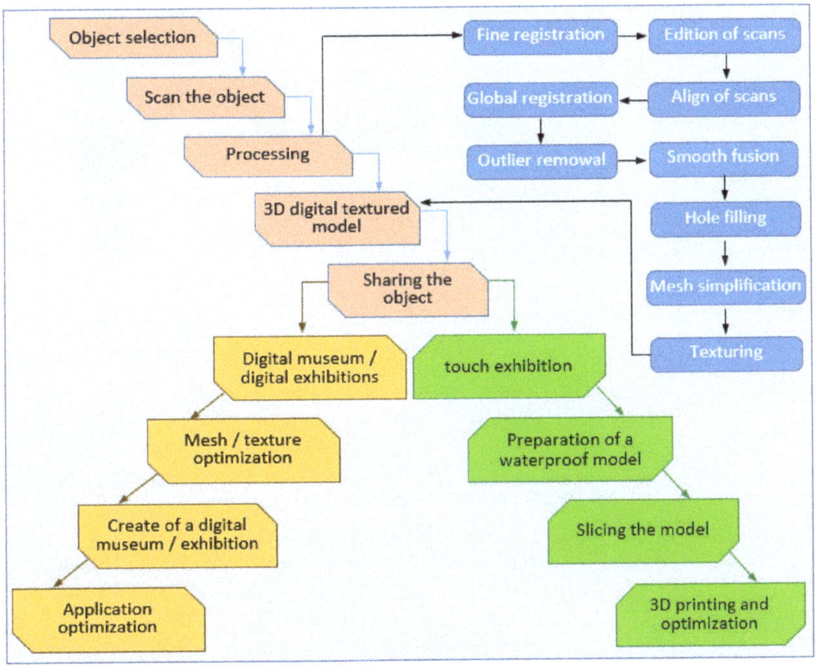

Figure 2. Procedure for making museum objects available.

The developed procedure proposes two paths for presenting museum objects: (1) digital sharing on websites or creating virtual museums, and (2) 3D printing, which makes it possible to share models with visually impaired people. Each stage is based on knowledge and experience in the field of museology and IT. Some activities are common to both paths, which makes the developed methodology universal.

3.3. Description of Objects

The Afrasiab Museum (Samarkand, Uzbekistan) in its collection has many interesting historical exhibits from the ancient city of Afrasiyab. The city was located in the commercial areas of the Silk Road through which goods were transported from China to Europe. It developed around the 5th century BC and was destroyed in 1220 by the invasion of Genghis Khan [36]. Selected exhibits of this museum were the subject of research, analysis and consideration of various scientific articles [37–39]. The work [37] presents a digital reconstruction of a damaged jug, ref. [38] presents an application that allows for the presentation of exhibits from the Afrasiyab Museum in VR, while [39] presents a comparative analysis for obtaining digital 3D models with the use of structured light scanning and the Structure from Motion method. Two objects were selected for the implementation of this study [40].

The first object is a damaged clay jug, about 21 cm in diameter and 35 cm high. This object has not been preserved in its entirety. Part of the side wall and the bottom are missing. The fragment of the vessel that has survived has a handle, a neck with a spout and a part of the side wall. The jug does not have any special decorations except for the ribbon, which is below the neck. It was used to transport and store water. The second object is a mug with a handle, also made of clay. The top of the handle is shaped like a ram's head. The vessel is 8 cm high and 10 cm in diameter. It has been preserved in its entirety and has no visible damage, as seen in Figure 3.

(a)

(b)

(c)

Figure 3. View of objects for digitisation: (**a**) mug; (**b**) jug; (**c**) work with the Artec Spider scanner.

3.4. Acquisition and Data Processing

Data acquisition was carried out during the first Scientific Expedition of the Lublin University of Technology to Central Asia in 2017 [41] and the University's fifth such expedition in 2021 [40] to museums in Uzbekistan where both objects were exhibited. The main problem of the 3D digitisation of objects was to build an appropriate place to scan using available furniture, providing access to power for the devices used, relatively good lighting for the artefacts and their safe extraction from showcases, which typically came from the end of the 20th century and were not adapted to opening.

Hand-held 3D scanners working with structured light technology were used to scan the objects. In the case of the jug, the Artec Spieder scanner was used, and for the digitisation of the mug, the Artec Eva. Both scanners allow for relatively fast execution of precise measurements in high resolution and, importantly, do not require sticking positioning markers on the digitised object, which is particularly important when scanning museum objects in in situ conditions. The used structured light is safe for the user and non-invasive for museum artefacts [10,42]. The basic data of these scanners are presented in Table 1.

Table 1. Comparison of the basic parameters of scanners Artec Eva and Artec Spider.

Parameters	Artec Eva	Artec Spider
Texture capture capability	Yes	Yes
Scanning technology	Structured white light	Flashing light bulb, blue light (no laser)
3D resolution, max.	0.5 mm	0.1 mm
3D point accuracy, up to	0.1 mm	0.05 mm
Linear field of view	214 × 148 mm	90 × 70 mm
Height × Width	Up to 536 × 371 mm	up 180 × 140 mm
Scanner distance from object	0.4–1 m	0.17–0.35 m
Video frames, up to	16 fps	7.5 fps

Table 1. *Cont.*

Parameters	Artec Eva	Artec Spider
Record output formats	OBJ, PLY, WRL, STL, AOP, ASCII, PTX, E57, XYZRGB	OBJ, PLY, WRL, STL, AOP, ASCII, PTX, E57, XYZRGB
Weight	0.85 kg	0.85 kg
Interface	1 × USB 2.0 or USB 3.0	1 × USB 2.0 or USB 3.0
Supported Windows systems	Windows 7, 8, 10—x64	Windows 7, 8, 10—x64
Minimum computer requirements	I5 (recommended I7), 8–12 GB RAM, NVIDIA GeForce 2GB	I5 (recommended I7), 8–12 GB RAM, NVIDIA GeForce 2GB

The scanning of the objects in Figure 3 was carried out in in situ conditions. For data acquisition during scanning, a software dedicated to the Artec Studio Professional scanner and a laptop with a 4-core Intel Core i5 processor and 16 GB of RAM were used. Eight scans were made for the jug-type object. This allowed for the precise acquisition of data about the facility. The data file in the form of a point cloud was approx. 5 GB in size. Six scans were taken for the mug and a 1.35 GB file was obtained. The processing of the obtained data was carried out according to the procedure presented in Figure 4. The same software as for data acquisition was used—Artec Studio Professional, as seen in Figure 4b–d.

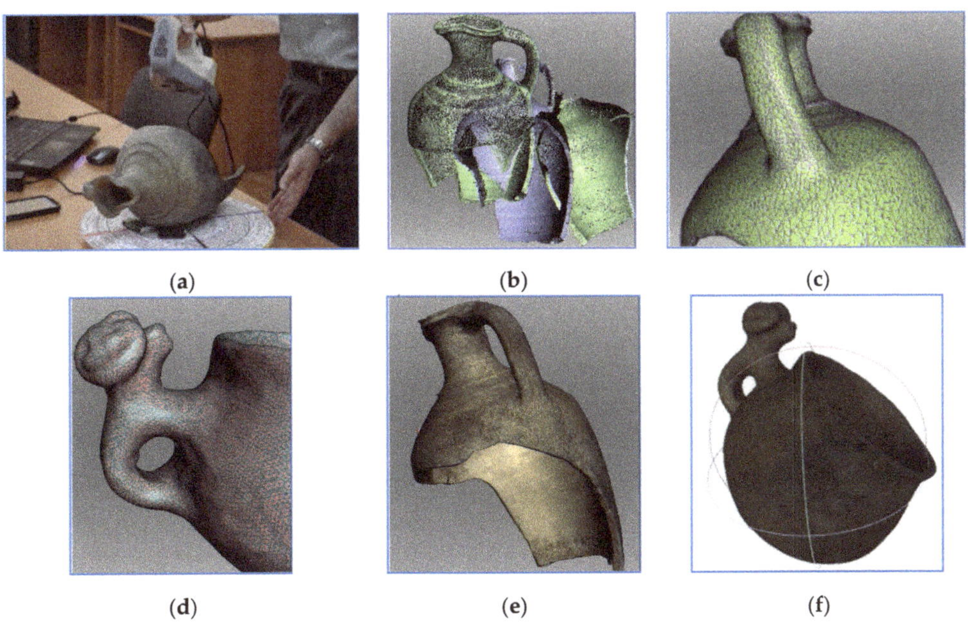

Figure 4. View of: (**a**) the scanning process; (**b**) data processing from the scanning; (**c**,**d**) the 3D mesh model; (**e**,**f**) the textured 3D mesh model.

Processing was carried out using a desktop computer with an Intel Core i7 processor, 48 GB of RAM and an NVIDIA Quadro K2200 graphics card with built-in 4 GB GDDR5 memory. The use of equipment with high computing power allowed for efficient data processing activities. As a result, mesh digital 3D models with the texture applied were obtained (Figure 4c–f). The models were exported as files in the .obj format, and textures as .jpg files. The sizes of the obtained files after optimisation were as follows: for the mug model—9.65 MB, for the jug—12.7 MB.

During the data processing, the following stages were followed:

(i) Registration 1—fine registration, introduction of single scans stored in local coordinate systems;

(ii) Edition of scans—removed artefacts, used selection tools such as: Lasso selection, Cutoff-plane selection and 3D selection;
(iii) Alignment of scans—scans were adjusted by specifying 3–4 pairs of characteristic points on the scans;
(iv) Registration 2—global registration, transfer of single scans from local coordinate systems into the global system, Figure 5a,b;
(v) Fusion 1—outlier removal, object surface points acquired during scanning that are too far from other acquired points are deleted;
(vi) Fusion 2—smooth fusion, assembling the partial scans into one object in the new coordinate system, generating a mesh model;
(vii) Postprocessing 1—if there were holes on the surface of the digital model, they were repaired by using the hole filing or fix holes tools;
(viii) Postprocessing 2—mesh simplification, optimization of the number of faces in terms of the complexity of the digital model and the size of the final file;
(ix) Applying texture—mapping the texture obtained when scanning an object to obtain a photorealistic digital model.

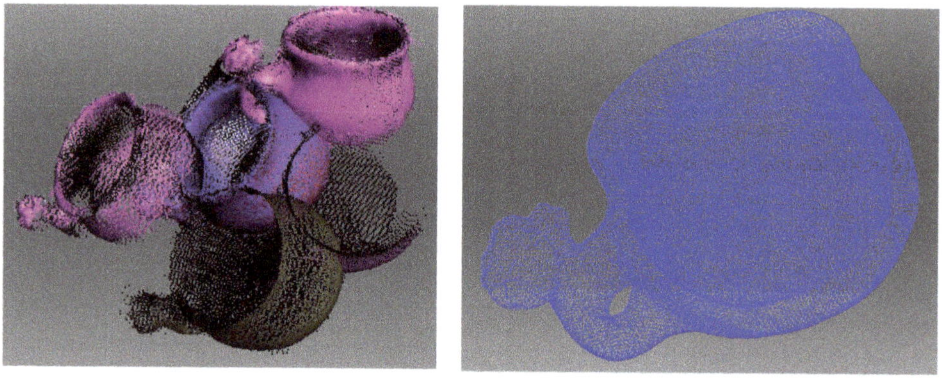

(a) (b)

Figure 5. View: (**a**) single scans in local coordinate systems; (**b**) scans in the global coordinate system.

3.5. Preparation of 3D Models for Printing

According to Figure 2, digital 3D models can be made available in digital or 3D printed form. The authors of the article have so far focused mainly on the first method of disseminating digital 3D models of museum objects. The objects were made available by publishing them: on the Internet as virtual 3D exhibitions [40], in the form of mobile applications for the presentation of 3D models [43], as a virtual museum [44] and as applications in the VR version [45,46]. The authors presented 3D models for visually impaired people only in [23].

In this study, copies of museum objects (the mug and jug) were made for the purpose of testing the suitability of such replicas for visually impaired people. The FFF printing technology was used due to the widespread availability of printers with an appropriate size of working space.

In the first stage of the works (Figure 2), the activities were related to the preparation of the so-called digital waterproof models, because only such models can be directed to 3D printing. The post-processing performed allowed for the closing of all surfaces and the removal of unnecessary edges and protruding surfaces. This task was performed with the use of 3D models generated in the previous stages, saved in the .obj format, and a set of tools for the free Blender program called 3D-Print Toolbox. 3D-Print Toolbox has specialised tools designed to identify surface geometry errors that may cause problems in the printing process. The watertight models prepared (using the "Slicer" type program) were transformed into layered models (Figure 6a) and supports were added in the required places. In order to ensure the stiffness of the object, a filling of the "grid" type was used. The printing parameters shown in Table 2 were adopted for both models. Wall Line Count was set to 3 to improve the quality of the outer surface to be touched. In order to verify the effect of the height of the layers on the roughness of the surfaces palpable by visually impaired persons, the thickness of the layer was 0.12 mm for the mug and 0.25 for the jug.

(a) (b) (c)

Figure 6. View of the objects: (**a**) layered model of the jug ready for printing (25% of the height); (**b**) a printed replica of the jug with supports; (**c**) a printed mug after cleaning it off supports.

Table 2. The printing parameters.

Print Parameters	Value
Wall thickness	0.6 mm
Wall line count	3 uL
Top/bottom thickness	0.84 mm
Printing temperature	210 °C
Build plate temperature	60 °C
Print speed	50 mm/s
Fan speed	100%
Support overhang angle	39°

The 1:1 scale was used to print the mug model, while for the jug the 1:2 and 1:1 scales were chosen. Three-dimensional printing was performed on a Makerbot replicator Z18 printer (Figure 6b), by using PLA-type plastic with a diameter of 1.75 mm. Printing the mug took 26 h, and 122 g (41 m) of material was used. Printing the jug in the 1:2 scale took 25 h, using 200 g (68 m running) of plastic, and in the 1:1 scale printing took 102 h (over 820 g of plastic with a wire length of almost 250 m were used). Finally, the printed models were cleaned off the supports (Figure 6c).

3.6. Model with Braille Text

The authors also prepared an experimental digital 3D model of the jug on the surface of which information was written using Braille. These actions were performed in Blender and the followed procedure included the following steps:

A. Preparation of letters in the Braille alphabet in 1:1 scale;
B. Composing a single word from the letters on the plate;
C. Scaling the object (jug) to the size in which it will be printed;
D. Incorporating tiles (in 1:1 scale) with words into the surface of the object;
E. Merging objects into one whole;
F. Generating the object for printing.

Stage A. Prepared single letters in Braille are 3D objects represented by mesh models. All the models were saved as separate files in the .obj format, with no texture, as seen in Figure 7.

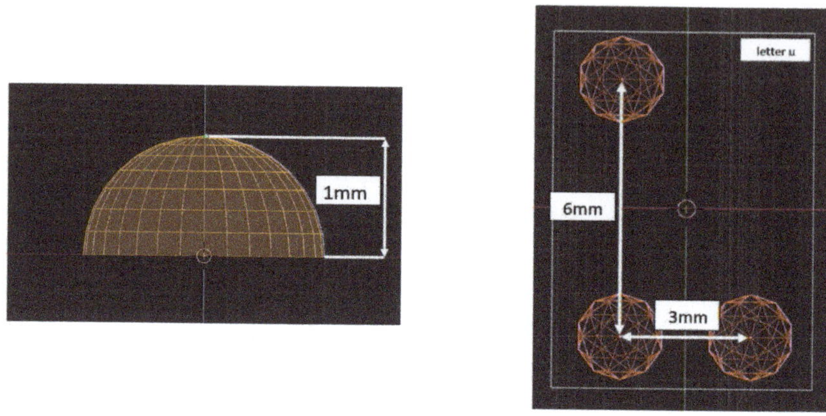

(**a**) (**b**)

Figure 7. Preview of the preparation of letters in Blender and dimensioning: (**a**) a single element for creating letters; (**b**) letter "u".

Stage B. Word generation was preceded by designing a plate slightly larger than the height of the 6-point Braille letter, 2 mm thick and 10 letters long. Creating words consisted in inserting subsequent letters as ready files on the designed plate. If the word was shorter than the original length of the plate, the unnecessary length was cut off; otherwise, it was lengthened. The individual words were saved as solid-type objects in separate files in the .obj format without texture, as seen in Figure 8.

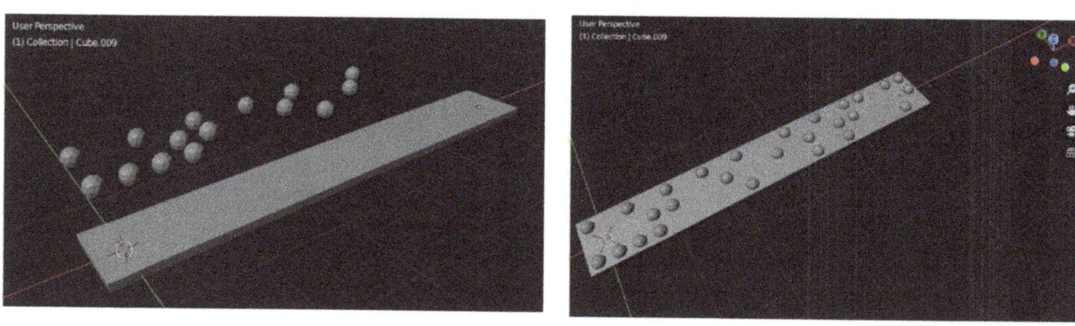

(**a**) (**b**)

Figure 8. Preview of text preparation in Blender—Stage B: (**a**) creating inscriptions in Braille; (**b**) modelling tiles with text in Braille (the text in Polish—Uzbekistan, in ang. Uzbekistan).

Stage C. The existing standards defining the sizes of individual cones (elements) that make up a letter and the distance between them mean that the object with the inscriptions cannot be scaled, because reduced letters may become unreadable for a blind person. Hence, a decision should be made on the final size of the object and its proper scaling.

Stage D. Composing the inscriptions with the tiles consists in importing the relevant files and their appropriate placement on the surface of the object. The matter is rather simple when dealing with flat surfaces. In the case of a jug, the outer surfaces form hypersurfaces, so the tiles with the words will stick out more in some places than in others, as seen in Figure 9.

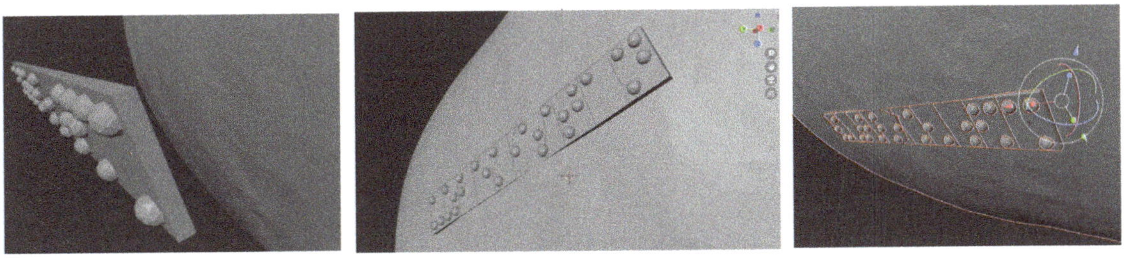

(a) (b) (c)

Figure 9. Preview of inserting inscriptions into the surface of the object in Blender—Stage D: (**a**) placing the inscription above the surface, (**b**) introducing the inscription into the surface of the object, (**c**) merging the objects into a whole—Stage E.

Stage E. Integrating the object into a whole consists in performing a "Union" type Boolean operation on the object model and individual tiles with words, as seen in Figure 9c.

Stage F. The merged model of the object with the inscriptions in Braille was imported into a Slicer program that generated a model for printing with the necessary supports. An important issue is to place the model in the replicator space in such a way that the model supports do not imprint on the inscriptions, because removing them may damage the inscriptions, as seen in Figure 10. The authors chose the upside-down position due to the generation of fewer supports.

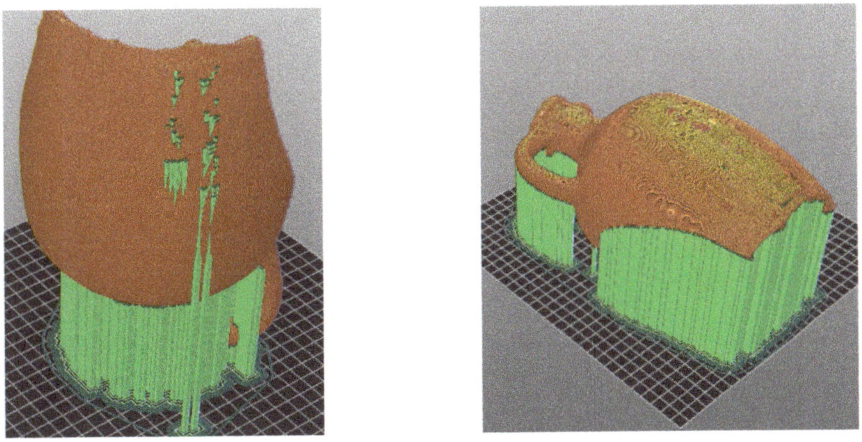

(a) (b)

Figure 10. Preview of alternative ways of placing the object in the workspace, simulation in Slicer—Stage F (the green colour represents the supports): (**a**) placing the object in an inverted position upright; (**b**) placing the object in a horizontal position with the inscriptions facing up.

Due to the length of the 1:1 scale printing process of the jug, it was decided to use a version of the model representing 80% of its actual size. Printing time was reduced to approximately 70 h. For the pilot study, the inscriptions in the following lines were scaled to verify whether the use of 3D printing with FFF technology with the selected printing parameters (Table 1) would not introduce some inaccuracies that would make it impossible for a blind person to read the words, as seen in Figure 11.

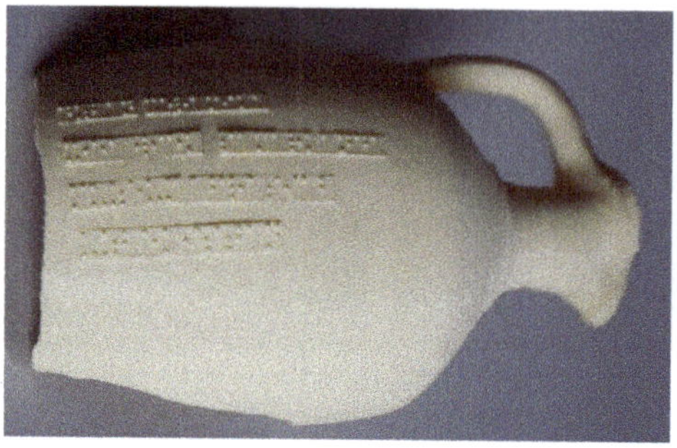

(a)

(b)

Figure 11. (**a**) A model of a jug with integrated descriptions in Braille; (**b**) the scales of the inscriptions in the following lines (starting from the top) were: 100%, 110%, 120% and 130%.

The following texts (of course in Polish) have been written in the following lines:

- First line: dzban ceramika (ang. pottery jug);
- Second line: XII wiek samarkanda (ang. 12th century Samarkand);
- Third line: muzeum afrasiab (ang. the Afrasiab museum);
- Fourth line: uzbekistan (ang. Uzbekistan).

It should be remembered that Braille only has lowercase letters.

4. Pilot Studies

The pilot studies were carried out in two stages. The first stage consisted in examining four people (two women and two men aged 45–55 years) with simulated pattern dysfunction. In the second stage of the study, one man was completely blind from birth (age group 65+) and had been using Braille books since childhood. All persons consented to the disclosure of their image.

The aim of the research carried out on a group of people with simulated visual impairment was to test the acceptance of 3D printed models for kinaesthetic cognition. Information was collected about the perceived quality of the surface of the object as a whole and about the places of the object where the supports were printed, as well as about the size of the objects and the recognition of its individual details. The research was carried out in "Lab 3D" at the Lublin University of Technology.

In the case of the blind person, the scope of the research was extended by collecting information on the legibility of the descriptions written in Braille (in various scales, as seen in Figure 11b), which were integrated onto the surface of the object. The research was conducted at the respondent's place of residence.

Stage 1—tests with simulated visual impairment.

The research was carried out according to the following scheme:

- Participants were brought into the room and had a non-translucent eye shield;
- The objects were placed in front of the examined person;
- The participants of the study were asked to try to classify the object into the appropriate group (e.g., mug, jug shell) and to identify individual details of the touched object (handle, spout condition, decorative elements, Braille inscriptions);
- The participants examined the surface roughness of the individual copy elements, as seen in Figure 12;
- The participants were asked to comment on the size of the tested object (this only concerned the jug).

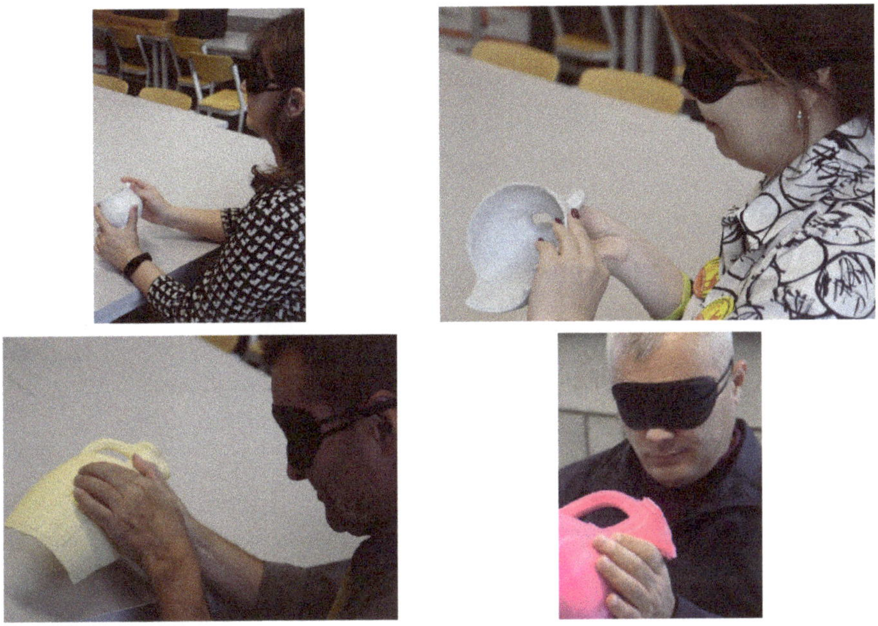

Figure 12. Studies involving people with simulated visual impairment.

Stage 2—examination of a blind person.

The research concerned the same scope as the research carried out in Stage 1. The recognition and legibility of the texts written in Braille were also examined, as seen in Figure 13.

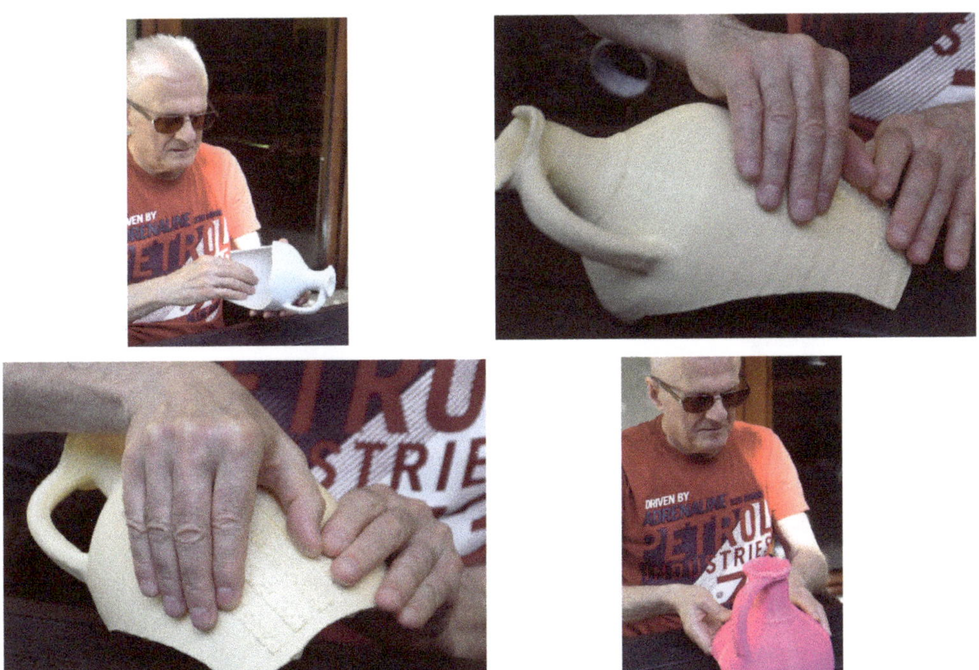

Figure 13. Research with the participation of a blind person.

The subjects were asked to comment on their impressions and what they felt through touch out loud. The entire course of the research was recorded on a Dictaphone (the respondents gave their consent). This allowed for their later listening, study and analysis. This solution allowed them to focus on conducting the research and did not distract the respondents. Before starting the research, the participants were informed that they would receive two objects for recognition. There was no information about the technique with which they were made and where they came from. Information was also provided on what they should pay attention to when touching the objects. During the research, the person conducting the experiment asked additional questions and asked about some details of the object. The order in which the objects were given was random.

5. Research Results and Discussion

The research was qualitative due to the small size of the research group. The main results of the research carried out in both stages are summarised in Table 3. This presentation of the results made it easy to compare them.

Table 3. Answers obtained to research questions.

No.	Research Element	People with Simulated Visual Impairment	A Person Who Was Blind from Birth
1.	Object recognition	+	+
2.	Was the cup roughness correct?	The surface was smooth and did not raise any objections.	Identical assessment.
3.	Were the details of the details of the cup recognisable?	Subjects felt a decorative feature on the top of the handle, but it was not recognised.	Identical assessment.
4.	Was the smoothness/roughness of the jug appropriate?	• The entire surface of the 1:2 scale model of the jug was appropriate. • The surface of the 1:1 scale model of the jug was appropriate in the part where there were no supports. • The roughness at the spout was too high and thus unacceptable. • The 80% size model area was appropriate.	• Identical assessment. • Identical assessment. • Identical assessment. • Identical assessment.
5.	Details detected in the jug model.	• The handle and the spout were recognised. • The tiles integrated into the surface were not recognised.	• The handle and the spout were recognised. • Braille text tiles were recognised.
6.	Was the text in Braille legible?	Not applicable. People did not know this alphabet.	• All texts were legible. • Enlarged texts (scaled to 110%, 120% and 130%) were not more readable than standard size text. • Letter size magnification was not needed.
7.	Did the effect of the size of the object matter (for the jug)?	The size of the jug did not matter when recognised by touch.	Identical assessment.

The results obtained from the pilot studies allow for the evaluation of the prepared copies of 3D models with FFF technology.

- Recognising the decorative element on the handle of the cup (ram's head) was too difficult. The blind person stated that even the enlargement of the cup together with this element would not contribute to its recognition, as they had never touched the head of a real ram;

- The blind person finding texts with inscriptions in Braille was a pleasant surprise, it had not been expected;
- Not all texts were immediately readable, as there were cases where the distances between individual letters were incorrect, i.e., too small. This concerned the distance between the letters "s" and "m" in the word "Samarkand". The respondent stated that the intervals should be larger. In another case, a printing defect appeared, which made it difficult to recognise the letters "c" and "e" in the word "ceramics". A similar situation was encountered with the inscription "museum"—there was a little difficulty when reading this word. There was no problem with reading the word "Afrasiab" (on the same line and scale) even though it was a proper name and the respondent did not understand the word. The word "Uzbekistan" was clearly legible;
- The respondents positively assessed the PLA material used for 3D printing, unequivocally stating that the material was pleasant to the touch and the analysed exhibits were light and could be freely manipulated.

It should be stated that at both stage 1 and 2 of the research, the objects were correctly recognised and named. When examining the cup, it was found that it could be a cup or a deep bowl with a handle. When examining the second object, it was stated that it was some kind of damaged jug (the shape was recognised thanks to the handle and characteristic spout).

6. Summary and Conclusions

Based on the technical work performed and the research carried out, the following conclusions can be drawn:

1. The methodology of the solutions proposed in this article, leading to the creation of a copy of a museum exhibit with a description in Braille integrated onto its surface, turned out to be effective.
2. The selection of digitisation, modelling and 3D printing technology, along with the software used for scanning, modelling, post-processing and preparation of the layered model for replication turned out to be appropriate and made it possible to effectively solve all emerging problems.
3. Incorporating descriptions of objects onto their surfaces using the Braille alphabet will contribute to multifunctional recognition—integration of knowing the shape of the exhibit through recognition by touch and the content written on its surface.
4. The procedure of preparing descriptions in Braille on separate plates makes it possible to pre-print them and verify both their correctness and legibility by a blind person. This solution will save time and costs when reprinting the entire 3D model if the descriptions are found to be defective.
5. The use of 3D printing technology in the fused filament fabrication version allowed for the production of light 3D models to be recognised by touch, which the respondents could freely manipulate. The results of the pilot studies carried out on a group of people with simulated visual impairment and on a person who is blind from birth indicate that these models are useful for sharing cultural heritage objects.
6. Printed museum models in the form of 3D replicas of these objects will allow them to be used in many ways. They can be exhibited in museums, printed at home, used for the training of conservators as well as used as archives.

7. Further Works

The authors intend to continue their work in two directions:

Developing a methodology for automating the creation of texts in Braille and their application on cylindrical surfaces of digital 3D models.

Generating 3D models with digital light processing technology using liquid-crystal displays (LCDs) for photopolymer curing and verifying the suitability of such models for the blind.

Author Contributions: Conceptualization, J.M. and M.B.; methodology, J.M., M.B. and S.K.; software, M.B. and S.K.; validation, J.M., M.B. and S.K.; formal analysis, J.M. and M.B.; investigation, J.M., M.B. and S.K.; resources, J.M. and M.B.; data curation, J.M., M.B. and S.K.; writing—original draft preparation, J.M., M.B. and S.K.; writing—review and editing, J.M., M.B. and S.K.; visualization, J.M., M.B. and S.K.; supervision, J.M.; project administration, J.M.; funding acquisition, J.M., M.B. and S.K. All authors have read and agreed to the published version of the manuscript.

Funding: The work was co-financed by the Lublin University of Technology Scientific Fund FD-20/IT (Technical Information Technology and Telecommunications).

Institutional Review Board Statement: This research was carried out in accordance with the decision No. 5/2020 of the Committee for Ethics of Scientific Research of the Lublin University of Technology of 15 July 2020 on the research project: Research on the usability of the created 3D museum exhibits for kinaesthetic cognition.

Informed Consent Statement: Informed consent was obtained from all subjects involved in the study.

Data Availability Statement: Data on digital 3D models are stored by M.B. and can be made available after individual notification by email: m.barszcz@pollub.pl.

Conflicts of Interest: The authors declare no conflict of interest.

References

1. Elmannai, W.; Elleithy, K. Sensor-Based Assistive Devices for Visually-Impaired People: Current Status, Challenges, and Future Directions. *Sensors MDPI* **2017**, *17*, 565. [CrossRef] [PubMed]
2. Pillow, B.; Banks, M.; Reeves, D. Equal Access for All: Providing for Impaired Stakeholders in a Museum Setting. Available online: http://www.byronpillow.com/uploads/7/6/7/9/76797585/disability_access.pdf (accessed on 18 November 2021).
3. Skalska-Cimer, B.; Kudłaczka, A. Virtual Museum. Museum of the Future. *Tech. Trans.* **2022**, *4*, e2022004. [CrossRef] [PubMed]
4. Leporini, B.; Norscia, I. *Translating Museum Visual Contents into Descriptions for Blind Users: A Multidisciplinary Approach. Computer-Aided Design of User Interfaces VI*; Springer: London, UK, 2009; pp. 47–57.
5. Petrie, H.; King, N.; Weisen, M. The Accessibility of Museum Web Sites: Results from an English Investigation and International Comparisons. In *Museums and the Web 2005*; Trant, J., Bearman, D., Eds.; Archives & Museum Informatics: Toronto, ON, Canada, 2005.
6. Smithsonian Guidelines for Accessible Exhibition Design. Available online: https://www.si.edu/Accessibility/SGAED (accessed on 25 January 2022).
7. Williams, T.L. More Than just a Novelty? Museum Visitor Interactions with 3D Printed Artifacts. Master's Thesis, University of Washington, Washington, DC, USA, 2017.
8. Ebrahim, M.A.-B. 3D Laser Scanners'. *Int. J. Sci. Res* **2015**, *4*, 5–611.
9. Díaz, T. Un nuevo espacio expositivo dentro del Museo Tiflológico: La sordoceguera. Antecedentes históricos. RdM. Revista de Museología: Publicación científica al servicio de la comunidad museológica. *Dialnet* **2019**, *74*, 111–118.
10. Adams, J.W.; Olah, A.; McCurry, M.R.; Potze, S. Surface model and tomographic archive of fossil primate and other mammal holotype and paratype specimens of the ditsong National Museum of Natural History, Pretoria, South Africa. *PLoS ONE* **2015**, *10*, e0139800. [CrossRef]
11. Stoddart, S.; Caroline, A. The British Museum. *Antiquity* **2002**, *76*, 1063.
12. Segal, H.; Jacobs, W. Computers and Museums: Problems and opportunities of Display and Interpretation. *Am. Q.* **1990**, *42*, 637. [CrossRef]
13. Luximona, A.; Luximonb, Y. New Technologies—3D Scanning, 3D Design, and 3D Printing. In *Handbook of Footwear Design and Manufacture*, 2nd ed.; Woodhead Publishing: Sawston, UK, 2021; pp. 477–503.
14. Ada, N.; Pirnar, I.; Huseyin Altin, O. Vitality of Strategic Museum Management: An Application from Turkish Museums. *Alanya Akad. Bakis* **2022**, *6*, 1891–1905. [CrossRef]
15. Museum Van Loon Portal. Available online: https://www.museumvanloon.nl/ (accessed on 18 January 2022).
16. Merriman, N. University museums: Problems, policy and progres. *Archeaology Int.* **2001**, *5*, 57–59.
17. Self-Guided Museum Visits for Visually Impaired People. Available online: https://www.rehacare.com/en/Interviews/We_asked_...../Older_We_asked_interviews/Self-guided_museum_visits_for_visually_impaired_people (accessed on 18 November 2021).
18. D'Agnano, F.; Balletti, C.; Guerra, F.; Vernier, P. TOOTEKO: A case study of augmented reality for an accessible cultural heritage. Digitization, 3D printing and sensors for an audio-tactile experience. In Proceedings of the International Archives of the Photogrammetry, Remote Sensing and Spatial Information Sciences, 3D Virtual Reconstruction and Visualization of Complex Architectures, Avila, Spain, 25–27 February 2015; pp. 207–213.
19. Reichinger, A.; Schröder, S.; Löw, C.; Sportun, S.; Reichl, P.; Purgathofer, W. Spaghetti, Sink and Sarcophagus: Design Explorations of Tactile Artworks for Visually Impaired People, NordiCHI '16. In Proceedings of the 9th Nordic Conference on Human-Computer Interaction, Gothenburg, Sweden, 23–27 October 2016; Volume 82, pp. 1–6.

20. Neumüller, M.; Reichinger, A.; Rist, F.; Kern, C. 3D Printing for Cultural Heritage: Preservation, Accessibility, Research and Education. In *3D Research Challenges in Cultural Heritage, A Roadmap in Digital Heritage Preservation*; Ioannides, M., Quak, E., Eds.; Springer: Berlin/Heidelberg, Germany, 2014; pp. 119–134.
21. Reichinger, A.; Fuhrmann, A.; Maierhofer, S.; Purgathofer, W. Gesture-Based Interactive Audio Guide on Tactile Reliefs. In Proceedings of the ASSETS '16. 18th International ACM SIGACCESS Conference on Computers and Accessibility, Reno, NV, USA, 23–26 October 2016; pp. 91–100.
22. Povroznik, N. 3D Models of Ancient Greek Collection of the Perm University History Museum. In Proceedings of the Digital Cultural Heritage, Final Conference of the Marie Skłodowska-Curie Initial Training Network for Digital Cultural Heritage, ITN-DCH, Olimje, Slovenia, 23–25 May 2017; pp. 144–154.
23. Montusiewicz, J.; Miłosz, M.; Kęsik, J. Technical aspects of museum exposition for visually impaired preparation using modern 3D technologies. In Proceedings of the 2018 IEEE Global Engineering Education Conference (EDUCON), Santa Cruz de Tenerife, Canary Islands, Spain, 17–20 April 2018; pp. 774–779.
24. Lazna, R.; Barvir, R.; Vondrakova, A.; Brus, J. Creating a Haptic 3D Model of Wenceslas Hill in Olomouc. *Appl. Sci.* **2022**, *12*, 10817. [CrossRef]
25. Akça, D.; Grün, A.; Breuckmann, B.; Lahanier, C. High Definition 3D-Scanning of Arts Objects and Paintings. In Proceedings of the Optical 3-D Measurement VIII, Zurich, Switzerland, 9–12 July 2007; Volume II, pp. 50–58.
26. Montusiewicz, J.; Czyż, Z.; Kayumov, R. Selected methods of making three-dimensional virtual models of museum ceramic objects. *Appl. Comput. Sci.* **2015**, *11*, 51–65.
27. Tucci, G.; Bonora, V.; Conti, A.; Fiorini, L. High-Quality 3d Models and Their Use in a Cultural Heritage Conservation Project. In Proceedings of the International Archives of the Photogrammetry Remote Sensing and Spatial Information Sciences Volume XLII-2/W5 26th International CIPA Symposium, Ottawa, ON, Canada, 28 August–1 September 2017.
28. Kowalski, M.; Naruniec, J.; Daniluk, M. LiveScan3D: A Fast and Inexpensive 3D Data Acquisition System for Multiple Kinect v2 Sensors. In Proceedings of the 2015 International Conference on 3D Vision, Lyon, France, 19–22 October 2015.
29. Daneshmand, M.; Helmi, A.; Avots, E.; Noroozi, F.; Alisinanoglu, F.; Arslan, H.S.; Gorbova, J.; Haamer, R.E.; Ozcinar, C.; Anbarjafari, G. 3D Scanning: A Comprehensive Survey. *arXiv* **2018**, arXiv:1801.08863v1.
30. Miłosz, M.; Montusiewicz, J.; Kęsik, J. *3D Information Technology in the Protection and Popularization of the Cultural Heritage of the Silk Road*; Wyd. Politechniki Lubelskiej: Lublin, Poland, 2022; p. 134. ISBN 978-83-7947-521-6. Available online: http://bc.pollub.pl/dlibra/publication/14114/edition/13771 (accessed on 23 August 2022).
31. Montusiewicz, J.; Barszcz, M.; Dziedzic, K.; Kęsik, J.; Miłosz, M.; Tokovarov, M. The Concept of a 3d Game Board to Recognise Architectural Monuments. In Proceedings of the 11th International Conference of Technology Education and Development, INTED, Valencia, Spain, 6–8 March 2017; pp. 8665–8674.
32. Bamidele, I.A. *Information Needs of Blind and Visually Impaired People*; Readers' Services Librarian, Laz Otti Memorial Library: Ilishan-Remom, Nigeria, 2019.
33. Maciuk, K.; Jakubiak, M.; Sylaiou, S.; Falk, J. Museums and the pandemic—How COVID-19 impacted museums as seen through the lens of the worlds' most visited art museums. *Int. J. Conserv. Sci.* **2022**, *2*, 609–618.
34. Montusiewicz, J.; Miłosz, M.; Kęsik, J.; Żyła, K. Structured-light 3D scanning of exhibited historical clothing—A first-ever methodical trial and its results. *Herit. Sci.* **2021**, *9*, 74. [CrossRef]
35. Kęsik, J.; Miłosz, M.; Montusiewicz, J.; Samarov, K. Documenting the geometry of large architectural monuments using 3D scanning—The case of the dome of the Golden Mosque of the Tillya-Kori Madrasah in Samarkand. *Digit. Appl. Archaeol. Cult. Herit.* **2021**, *22*, e00199. [CrossRef]
36. Fedorov-Davydov, G.A. Archaeological research in central Asia of the Muslim period. *World Archaeol.* **1983**, *14*, 393–405. [CrossRef]
37. Montusiewicz, J.; Barszcz, M.; Dziedzic, K. Photorealistic 3D digital reconstruction of a clay pitcher. *Adv. Sci. Technol. Res. J.* **2019**, *13*, 255–263. [CrossRef]
38. Miłosz, M.; Skulimowski, S.; Kęsik, J.; Montusiewicz, J. Virtual and interactive museum of archaeological artefacts from Afrasiyab—An ancient city on the silk road. *Digit. Appl. Archaeol. Cult. Herit.* **2020**, *18*, e00155.
39. Barszcz, M.; Montusiewicz, J.; Paśnikowska-Łukaszuk, M.; Sałamacha, A. Comparative Analysis of Digital Models of Objects of Cultural Heritage Obtained by the "3D SLS" and "SfM" Methods. *Appl. Sci.* **2021**, *11*, 5321. [CrossRef]
40. 3D Digital Silk Road Portal. Available online: https://silkroad3d.com/ (accessed on 23 August 2022).
41. Miłosz, M.; Montusiewicz, J.; Kęsik, J. 3D Information Technologies in Cultural Heritage Preservation and Popularization—A Series of Seminars for Museologists Made by Computer Scientists. In Proceedings of the EDULEARN 20, 12th Annual International Conference on Education and New Learning Technologies, Valencia, Spain, 6–7 July 2020; pp. 544–549.
42. Graciano, A.; Ortega, L.; Segura, R.J.; Feito, F.R. Digitization of religious artifacts with a structured light scanner. *Virtual Archaeol. Rev.* **2017**, *8*, 49–55. [CrossRef]
43. Skulimowski, S.; Badurowicz, M.; Barszcz, M.; Montusiewicz, J. Design and optimisation methods for interactive mobile VR visualisation. *IOP Conf. Ser. Mater. Sci. Eng.* **2019**, *710*, 012015. [CrossRef]
44. Ciekanowska, A.; Kiszczak-Gliński, A.; Dziedzic, K. Comparative analysis of Unity and Unreal Engine efficiency in creating virtual exhibitions of 3D scanned models. *J. Comput. Sci. Inst.* **2021**, *20*, 247–253. [CrossRef]

45. Montusiewicz, J.; Miłosz, M.; Kęsik, J.; Kayumov, R. Multidisciplinary Technologies for Creating Virtual Museums—A Case of Archaeological Museum Development. In Proceedings of the INTED 2018: 12th International Technology, Education and Development Conference, Valencia, Spain, 5–7 March 2018; pp. 326–336.
46. Żyła, K.; Montusiewicz, J.; Skulimowski, S.; Kayumov, R. VR Technologies as an extension to the museum exhibition: A case study of the Silk Road museums in Samarkand. *Muzeol. Kult. Dedicstvo Museol. Cult. Herit.* **2020**, *4*, 73–93. [CrossRef]

applied sciences

MDPI

Article

Real-Time Rendering Engines Help Visualize, Model, and Animate Ancient Cities: An Example in Antioch

Kayhan Kaplan

Department of Digital Game Design, Bilgi University, Istanbul 34060, Turkey; kayhankaplan@hotmail.com

Abstract: This study aimed to revitalize ancient cities' history, culture, and social life, along with their physical environments, through computer-aided, three-dimensional (3D) images. This study will become an instrument for visualization in 3D ancient cities whose historical remains are not intact and are perhaps known only through books, maps, pictures, and inscriptions. Following historical investigations and comparative analyses of similar examples, objects or structures subject to change over centuries, partially damaged, or lost were visualized through the production of models that reflect their original historic design. The images obtained were converted to animations of 3d models with motion, sound, light, and related effects. In addition to animations, this study implemented another technique concerning cities: the Ancient Antioch—the modeled space was installed in a game engine to obtain simultaneous images. Despite certain limitations, the study findings, adequately based on scientific evidence, provide rich visual content for practical use in the three domains of presentation, research, and education. The hope is that this socio-historical, scientific, and esthetic study, established to model, visualize, and animate city life from approximately 1800–2000 years ago, will be instrumental to the further development of its mixed methodologies and provide a new synthetic process for an ongoing feedback loop. Perhaps the most important output of this study was the use of almost all of the images obtained in the Hatay Archaeology Museum. After the final stage of preparation, the presentation of Ancient Antioch is planned to be done in 2023 of March and opened to visitors on special platforms with Unity 3D. An actual excavation area's data can be constantly modeled and remodeled, thus improving the scientific feasibility, quality, and uniqueness of excavation work and the dissemination of scientific results.

Keywords: 3D reconstruction; game engine; digital archeology; virtual heritage; Antioch

Citation: Kaplan, K. Real-Time Rendering Engines Help Visualize, Model, and Animate Ancient Cities: An Example in Antioch. *Appl. Sci.* **2022**, *12*, 12316. https://doi.org/10.3390/app122312316

Academic Editors: Marek Milosz and Jacek Kęsik

Received: 20 September 2022
Accepted: 25 November 2022
Published: 1 December 2022

1. Introduction

This scientific, social, and esthetic study aimed to investigate cultural heritage by animating ancient cities' history, culture, and economic and social life, along with environmental factors such as topology and climate, through computer-aided, three-dimensional (3D) images. This is a means of safeguarding, conserving, and interpreting—without exploiting—ancient city sites whose major historical remains have not survived to the present but are understood to exist according to books, maps, scriptures, and pictures. From a comparative, theoretical, and practical examination of similar examples and historical research, the vision of this study is to manufacture and visualize models displaying the historical status of elements and structures subject to change, addition, and partial or complete destruction since their original design and construction. Even if this innovative, scientific, and obligatory study cannot completely reintegrate an artifact, computer-aided visualizations and restitution work can make it possible to explain a work's unique design, examine its historical development, and gain a better understanding of its remains. In other words, this study presents opportunities to understand historical and architectural connections within the entirety of a region and people aided by 3D presentations based on scientific, social, and historical information.

Rapid advancements in computer hardware and software are boosted by the increasing quality of computer graphics and animation processes. Therefore, interest in and the need for computer graphics has gradually increased and has begun to dominate in multiple fields, from daily life to diversified, specialized scientific branches. In fact, along with the formation of standards and the addition of new concepts, computer graphics have increasingly enhanced and enriched their subjects.

Of course, computer graphics are widely used in design and manufacturing, education and training, science and cognitive visualization, entertainment and advertisement, art and the display world, and cyberspace and augmented reality [1]. Currently, 3D content has become very popular not only because of the current technological trends (e.g., 3D films and television), but also because it provides better comprehension and appreciation of visual content [2].

Thanks to computer technologies, every place from the past, present, or future can be animated as a virtual model; modern virtual reconstruction is quite realistic because of improved computer systems and visualization peripherals and a better understanding and implementation of geometric and visual modeling techniques [3]. The virtual model is a valid cognitive tool and a fundamental medium through which a user (e.g., a scholar, student, or simply someone interested) can interact with 3D models and agents in a virtual environment. This technology can be applied to cultural heritage as a vessel for preservation, reconstruction, education, documentation, research, and dissemination [4]. Presenting structures or objects of cultural heritage in 3D models especially helps to deepen and innovate the perceptions of the past.

1.1. *Technological Materials: Recent Improvements*

Virtual realities created by modeling archeological objects and structures have been used in supporting archeology and the reconstruction of cultural heritage [5]. At the beginning of this century, however, information technology (IT) could only enable the modeling of structures and archeological objects of cultural heritage in animation form [6,7] Depending on the technology, computers and software have improved almost daily, with the final product's quality determined especially by the software used to produce a 3D model. Based on this development, computer graphic technology enables the modeling of structural groups and even cities in 3D. Today, 3D computer graphics are intended for designing groups of buildings, such as campuses and industrial complexes; planning cities; designing vehicles; and designing electronics, textiles, foods, and a myriad of other products.

The usage of computer-aided visualization is increasing daily, including but not limited to cultural heritage, virtual archaeology, 3D archeology, and museum simulation [8]. Museums are one of the leading institutions where cultural heritage products are stored, preserved, and exhibited. Due to the influence of the opportunities offered by developing technology, museums have started to accommodate their policies to the wishes of visitors with an understanding of contemporary museology.

Several opportunities have appeared to disseminate culture and heritage, thanks partly to the widespread use of information and communications technologies [9]. Traditionally, the way to appreciate tangible heritage was to visit museums and other cultural institutions in person. The use of virtual technologies and digital heritage is also relevant to the idea of a virtual museum, even more considering that the Internet offers new ways to spread and distribute information [10]. The virtual museum was originally developed to promote and reference regional museums where the pieces are located and encourage users to visit them. Virtual museums have become more important during the COVID-19 period. This can be seen in the fact that museums increased their online presence by 80% during the lockdown, offering more social media interactions, broadcasts, virtual tours, and online exhibitions, and observed more than a 40% increase in online visits [11]. Even with new adjustments to the virtual museum or interactive website, there was no major change regarding the interest in cultural heritage; people still tend to prefer to visit real-life museums than visual museums [9,12].

Computer-aided visualization practices have been used for a long time to ensure visitors' interaction with the works exhibited in the museum. The explanation of the objects in the museum is provided through simulations, reconstructions, object restoration, VR, AV, and computer-aided animations.

In recent years, the use of VR games in cultural heritage has been growing. VR games have increasingly found their way into museums and exhibitions, highlighting the increasing cultural value associated with games and the institutionalization of game culture [13]. The use of VR technology in museums not only increases multiple learning outcomes but also has a positive effect on the motivation and dynamic engagement of visitors. Particularly, VR and CAV in a real museum are more significant than a web-based museum [9].

Rua and Alvito used 3D models and virtual reality to reconstruct the archaeology and cultural heritage in the mansion of Casal de Freiria, located in Rome. The mansion, which consists of one building complex, is one of the first examples where 3D models are used in gaming engines. Considering technology's daily development, in the upcoming future, more complex and crowded scenes and models can be present in game engines [5].

The Unity game engine was used successfully for Hagia Sophia Church, located in İstanbul. Hagia Sophia was chosen as the sample study area of this study and was digitally restructured to be used in the project [14].

Another example of a game engine being used in museums is Labyrinth 3D. This application, primarily for educational and informational purposes, is based on 3D images of artworks in a museum and the interaction between them [15].

Rome Reborn (1997–2012) is a virtual ancient city built in an international effort involving research groups from the USA and Italy. The project is based on creating a digital model of ancient Rome as it appeared in late antiquity. The main purpose of the project is to spatialize and present theories about how the city looked at this point, which was the culmination of its development as the capital of the Roman Empire. The second purpose is to form a cyber framework where the model can be updated, expanded, and bettered. The Rome Reborn project distinguishes two types of elements in the model. Class I elements (detailed models) are sites that have detailed information about identification, location, and design. Approximately 30 elements fall into this category. Class II (procedural models) are the 6750 buildings and monuments (such as single-family houses, apartment buildings, warehouses, etc.) that lack precise information [16,17].

The ancient city of Palmyra—which is located in Syria—is geographically located close to the city of Antioch. With the help of the project Palmyra, 3D models of Colonnaded Street, the amphitheater, and the Bel and Baalshamin temples have been made. The models were not covered or textured but were instead left monochromatic [18,19]. Another project in Palmyra presents sketches and information given in diaries, along with archival photos, paintings, inscriptions, and sculptures, in an interactive web-based digital model of a monumental underground tomb, the so-called tomb of Hairan [20].

A study on the ruins of the ancient city was also carried out on the city of Karakorum (Mongolia). This project explores the first attempt to virtually rebuild the historic city of Karakorum based on ruins and archaeological findings from the 13th century AD. Only a group of buildings, called the Great Hall, was 3D modeled, and the platform was created with game engine integration [21].

Research on the virtual reconstruction of the historic city of Al-Zubarah in Qatar consisted of two phases. First, the creation of a volumetric solid model of the city and then the detailing of four important structures: the mosque, palace, towers, and bazaar [22].

Smith et al., (2019) tried to install cultural heritage, 3D modeling, and game engine technologies on a PC system. This article examines the development of a 3D model built from plans of a Roman edifice and its transfer into four game engines as vehicles for independent navigation around the 'virtual building' [23].

As a result, pilot studies are generally done in the form of 3D as modeling of several building blocks for the revival of cultural heritage in literature. Studies in which an ancient city is animated with its topography, landscape, and atmosphere have not been encountered, except for Rome Reborn 2.0. The most distinctive difference of our study is that an ancient city, which has very few ruins, is completely modeled with its topography and loaded into the game engine.

1.2. Technology–Comment–Reality: The London Charter

The rapid development of computer graphics and imaging provides modern archeologists with several tools to realistically model and visualize archeological sites in 3D. This, however, creates a tension between veridical and realistic modeling. Visually compelling models may lead people to falsely believe that there is very precise knowledge about the past appearance of a site [24]. Determining this fine line requires a set of principles on technology–comment reality.

The London Charter for the Computer-Based Visualization of Cultural Heritage (2.1) is a document written in 2009 by a pool of internationally renowned experts on digital technologies and virtual reality, which arose out of a recognition of the use of three-dimensional computer modeling in the historical, archaeological, and broader cultural heritage domain [25]. This Charter seeks to establish what is required for 3D visualization to be, and to be seen to be, as intellectually rigorous and robust as any other research method. The initiative has to be seen in the context of what has become a constant issue in 3D-visualization applications on cultural heritage: 'transparency' [26].

The transparency of sources and methods is very important for the validity and acceptance of research outputs. Thus, the outputs are not only dependent on the author's interpretation and are open to evaluation.

1.3. Archeological Materials: An Example in Antioch

Among the largest metropolises in the Roman Empire, Antioch, along with Alexandria and Constantinople, is regarded as an excellent city in times dominated by Hellenistic and Roman culture and a luxurious socioeconomic lifestyle [27]. Then, ranked as the third-largest city in the world, with a population of 300 to 400 thousand, Antioch became the capital of Roman Syria at the intersection between East and West [28]. Having magnificent structures, with its Hippodrome, arena, Imperial Palace, theaters, and Turkish baths, the city became the center of an empire, a social structure where the arts, performances, and Olympic Games were held.

Furthermore, Antioch did not lose its central characteristics until the great earthquake in 526, agreed to be the second most deadly natural disaster in history [29]. The great earthquake and several secondary earthquakes, Persian attacks, invasions, fires, and a great plague destroyed the city.

Because of earthquakes especially, most Hellenistic and Roman structures did not survive the ages. Consequently, information about Antioch's urbanism and architectural elements is obtained from written sources. Additionally called the Lost City, Antioch lies beneath land in the Antakya District, modern Hatay Province, situated in southernmost Turkey [30]. In the 1930s, Princeton University began the first archeological excavations of Antioch; however, very few Roman works were excavated.

2. Methodologies: Historical Research, Technological Adaptation, and Esthetics

Because this is a scientific, social, historical, and esthetic study, mixed methodologies are used, especially Kerlow's (2004) 3D computer-animation process [31]. This process is currently accepted and used in many commercial animated films, and here, it is further developed and adapted for this research process, which contains the following phases: data collection, analysis, modeling, coating, and rendering (Figure 1).

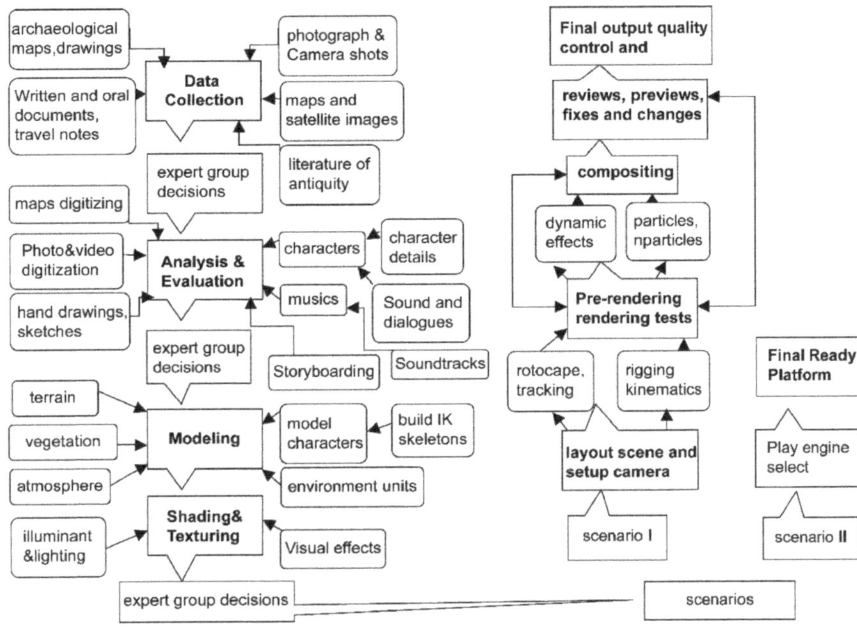

Figure 1. Visualization process for the ancient city of Antioch.

In addition, during the construction process of this study, the Rome Reborn project benefited significantly. As in the Rome Reborn Project, the buildings within the Ancient Antioch city are divided into two types. Class I elements (detailed models) are sites that have detailed information about identification, location, and design. Class II (procedural models) lack precise information, such as single-family houses, buildings, warehouses, etc. [16].

Two different scenarios are planned according to the assessment of the model's success. In Scenario I, drama and animation are produced; in models, scenes contain characters walking different routes and having different perspectives. In addition to the animation presentation, Scenario II writes the model city into a real-time game engine that receives rendering from the model synchronically. Thanks to this engine, the researcher, student, or interested user can wander along any route within the ancient city.

Materials

As the longest and most important research phase, modeling was conducted with data obtained from expert decisions. In this study, in addition to models manufactured in previous studies, to save time, detailed and ready models were obtained from several sources (for instance, Internet libraries and software). In the process, modeling begins with creating topography or terrain. Two-dimensional numerical data obtained via the ArcGIS 2.8, (CA, USA) and AutoCAD 2020 (CA, USA) software from satellite views and maps was transferred into the Vue 3.0 (Paris, France) and 3D MAX 2020 (CA, USA) modeling software and then dimensioned.

At this stage, structures such as houses, buildings, or warehouses, as well as rural areas such as vegetation and fields, about which there is little information, were procedurally modeled with the Vue ecosystem module.

Thus, a numerical height model of the entire Hatay region was obtained. During this phase, along with vegetation, landscaping, rivers, lakes, climatic, and other environmental elements reflecting the period, models of the city, particular buildings, roads, and other structural elements were modeled. Temples, arenas, a Hippodrome, a forum, an agora, and city palaces were processed in detail from drawings and other media with authenticity paramount.

At this phase, however, not having the information necessary for modeling structures limited the study's progress. Excavation notes provide information only pertaining to the planning and settling of structures; they fail to contain information about height, windows, roof cover, or any other details. Even extant textual information on a structure does not provide all the data necessary for a three-dimensional building model. Another modeling problem, for example, is that the sites of Ancient Antioch's important structures are buried under earth in present-day Antakya—the situation prevents excavation at the desired level. Thus, at this phase, an important theoretical compromise occurs for the sake of continuing research; the textual information on other ancient cities, which might represent presence, plan, and scale, but whose height, roof cover, windows, door, and column details are not available, was used for modeling structures. For the type of modeling used in this example, the following criteria were considered: belonging to the same period, having conventional Roman architectural characteristics, and, if possible, proximity to the Antioch region. For character modeling, research was conducted on pictures, and the drafts were obtained. Characters' clothing and equipment were modeled with diligence and aligned with expert advice. Nevertheless, even though every effort was made to base this study's version of Ancient Antioch on adequate scientific evidence, the model is still open to comment. In the final analysis, the study's visualized Antioch became a product of the project leader's point of view, and this scientific, socio-historical study was thus transformed into a study of esthetics and became a work of art.

In the study, camera tracking and camera points of view are determined according to the 'Scenario I' method for obtaining animation. In this phase, routes and targets of the camera were chosen. In addition, the camera speed was adjusted to the dynamics of moving objects.

Throughout the study, 'preliminary rendering' was obtained from settings whose light, shade, and coatings were completed from the predetermined camera tracking and points of view. In this process, with advice about the latest forms of animations, settings that disturb the viewer's eye or seem problematic in modeling or coating could be modified. Serving as a checkpoint, preliminary rendering was repeated for each camera point of view and moving object. Furthermore, character animations were created during this phase, and rendered images could be assessed.

Another method applied in Scenario II is the game engine. In rendering engines written in C++ computer language, all graphics must be written into the engine mathematically. After research into the necessary literature and previous studies, the decision was made to use an Object-Oriented Graphics Rendering Engine (OGRE) and Unity 3D. OGRE is an open-resource coded graphics processing and 3D game engine software group written and managed by a core team and supported by an increasing number of users. Unity 3D is the most common game engine used in the game and entertainment industry. The game engine's software characteristics enable the researcher, student, or interested user to reach any point within the ancient city and to take synchronic rendering images connected to the camera from any point.

3. Results for the Example of Ancient Antioch

For modeling Ancient Antioch, the Roman period (40–350 CE) was chosen. First, the structures to be modeled were chosen, and resource information was obtained. To this end, Downey (1961) [32] and Uggeri's (1996) [33] maps displaying Ancient Antioch were taken as a basis (Figure 2).

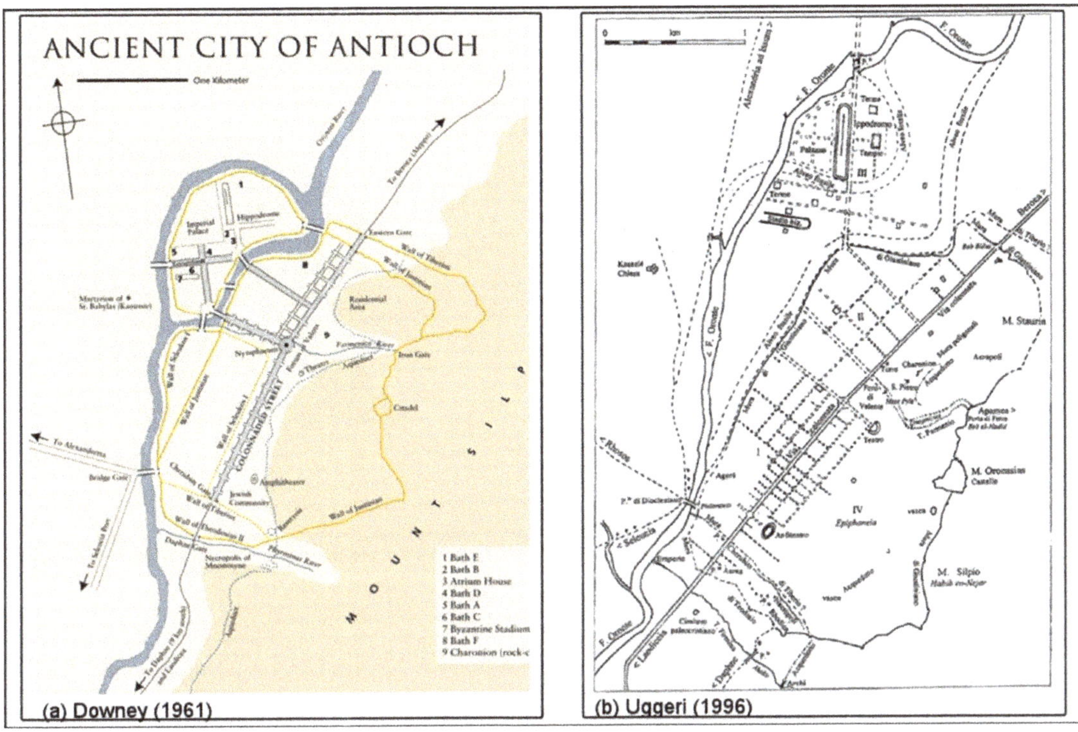

Figure 2. (**a**) Downey's (1961) map and (**b**) Uggeri's (1996) map of Ancient Antioch.

First, maps were started to be collected to create a 3D terrain model of Ancient Antioch. For this purpose, historical hand-drawn engravings, maps, and sketches were examined. The first topographical and cadastral studies in the region were started by the French in 1922. Those maps and updated 1/25,000 military topographic maps were compared and evaluated. In addition to these, historical and current satellite images were also used to ensure the accuracy of the data. As part of the Project, high resolution satellite images were used to support the general description of the ancient city under study and for the reconstruction of their topographies.

The results of this study include the creation of a composite historical database using ancient references and first topographical maps (1937 by France and later 1958), Landsat MSS (Multispectral Scanner System 1973, 1978, and 1988), Panchromatic Ikonos image (2018) data, and Sentinel1 (2018 and 2019) data. In this study, the data has been analyzed using the ArcMap 10.4 software.

The digital 2d maps were transferred to AutoCAD 2020 software. If there were deviations or distortions in the contour curves, they were checked and corrected. In this stage, AutoCAD software was used as a transmitter with a better 2D point control system. Later, these maps were converted into 3D terrain in 3D max 2020 software. Then, this terrain model was transferred to Vue 3.0 software and texture operations were performed (Figure 3). Vue 3.0 software is high photorealistic, and the animation controls are easy and fast., This software has been favored especially because it can quickly create crowded scenes as procedural with the ecosystem module.

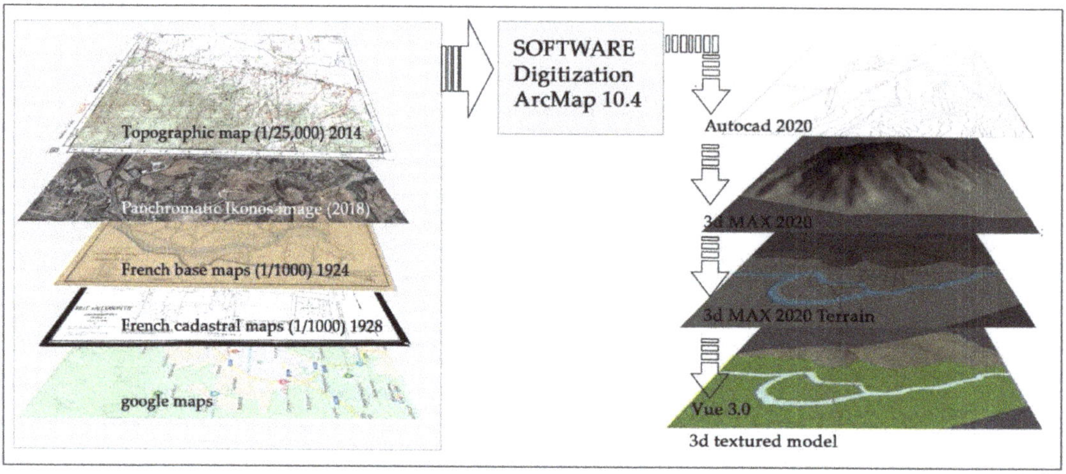

Figure 3. 3D terrain model construction stages of ancient Antioch.

The satellite images contribute little to the reconstruction of the Orontes River of the area of the Ancient Antioch; they merely show the current situation, with short water courses descending from the Sillipus mountain situated a little way to the East, perhaps following the same route as they did in ancient times. In the ancient city, we have more information about some important buildings, identity, location, and design information. Therefore, these structures were modeled in more detail. Of the structures in the city, the following were modeled: Colonnaded Street, the Imperial Palace, the Hippodrome, Nymphaeums, aqueducts, city walls, city gates, theater buildings, the amphitheater building, the stadium–arena, Tetrapol, the baths, and villas and houses (Figure 4).

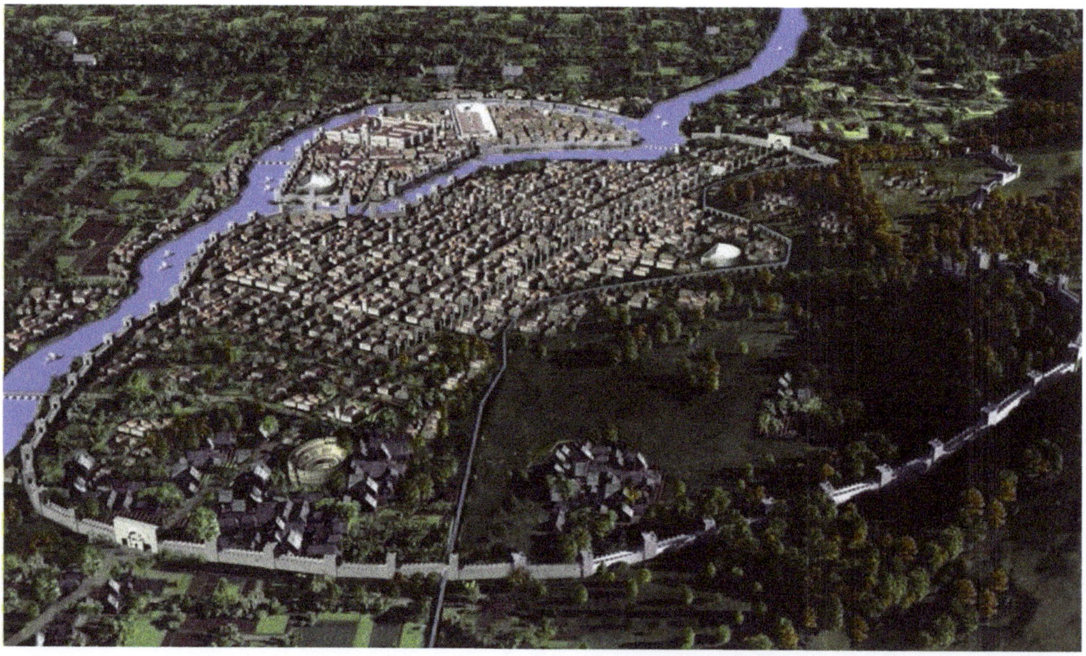

Figure 4. Model of Ancient Antioch (shown from South to North).

The most important structures of Antioch city (e.g., the palace, circus, stadium–arena, and the Tetrapol) are situated on an island formed by the Orontes (Asi) River (Figure 5).

Figure 5. Model of Ancient Antioch as situated along the Orontes River.

3.1. The Modeling of Ancient Antioch's Colonnaded Street

Like other large structures in the Antioch of the Roman period, no part of Colonnaded Street has survived to the present day. For modeling Colonnaded Street, excavation notes, historical records, and expert comments were used. First, the works of Lassus, who performed excavation in Antakya in the 1930s, formed a foundation. According to his excavation reports (1972) [34], Colonnaded Street, which is 3400 m long, is situated 11 m beneath Antakya's current Kurtulus Street.

As deduced from Downey's and Uggeri's maps and Lassus's excavation notes, Colonnaded Street does not lie along the city section called Seleucid. Instead, the street is situated at the edge of Mount Silpius between Daphne and Beroea Gates within Epiphania. Antioch's Colonnaded Street was maintained for eight centuries when Herod, Tiberius, Antonius Pius, and Justinian governed, but no structural changes were made. According to an expression of Libanius, the columns were lit in the evenings, and people would gather until late at night. At the same time, merchants hawked their ware at the sunder columned entrance [35].

Thus, Ancient Antioch's Colonnaded Street was modeled according to these descriptions. First, columns were constructed and situated according to Downey's map. Next, Lassus's excavation notes formed the basis for the road's width and length. Third, as mentioned by Lassus, buildings facing the street were modeled as shops. Finally, Roman architectural examples of the same period were used as designs for building facades (see Figure 6 for resource data [36,37].

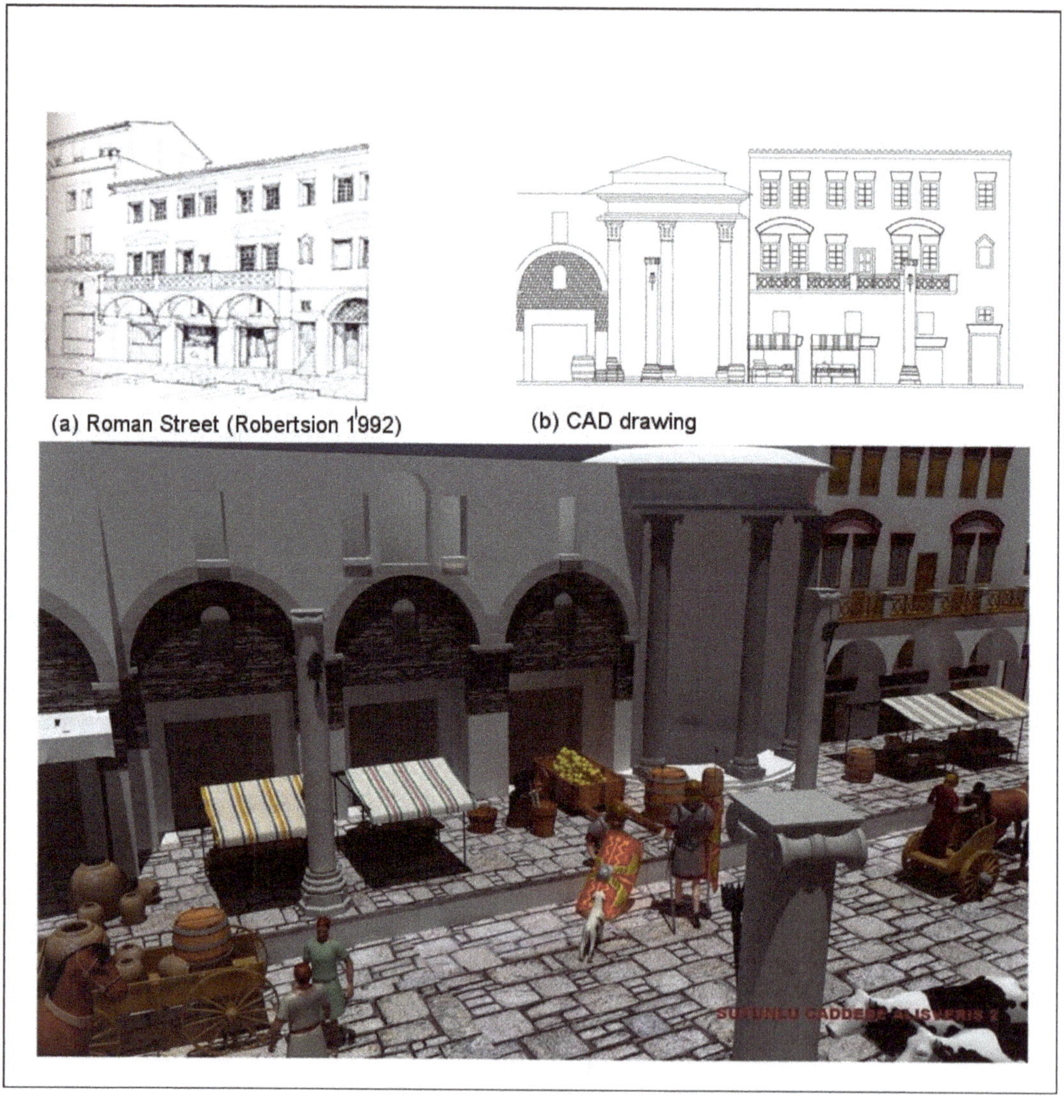

Figure 6. Shop buildings with Roman architecture of the ancient period, modeled for virtual Antioch, with data sources above.

3.2. The Modeling of Ancient Antioch's Imperial Palace

Malalas emphasized the high walls and complex structure of the Antioch Imperial Palace, which was constructed in 253 BC during the time of Gallienus. Similarly, Malalas also stated that King Trajan could see the Hippodrome from his chamber window in the Imperial Palace [38]. However, Libanus best described Antioch's Imperial Palace; on the island, the palace was a magnificent structure surrounded by high walls. Its northern part was open to the Orontes River. Its chambers and corridors were supported with columns. It became famous for its size and dimensions, similar to the Palatine Palace complex and the Circus Maximus structure in Rome. However, it was considered similar to the Diocletian Palace in Split [39].

For this study, these historians' expressions were used in terms of the shape and view for modeling the Antioch Imperial Palace. In addition, historical pictures and gravure paintings of Ancient Antioch provided information. The palace model was designed without detail and coating (Figure 7).

Figure 7. Ancient Antioch's Imperial Palace on the island in the Orontes River.

3.3. *The Modeling of Ancient Antioch's Hippodrome and Circus*

Antioch city is located on the island's center in Ancient Antakya; the Hippodrome is considered a part of the Imperial Palace complex, similar to the Maximus Hippodrome in Rome. As a hippodrome with a large capacity, this construction is thought to have accommodated 80,000 persons [40,41].

In the 1932 Antakya excavations by Princeton University, remains of the Hippodrome were found. Figure 8 shows a plan of the Hippodrome contained in excavation notes and the circus modeled according to this plan [40].

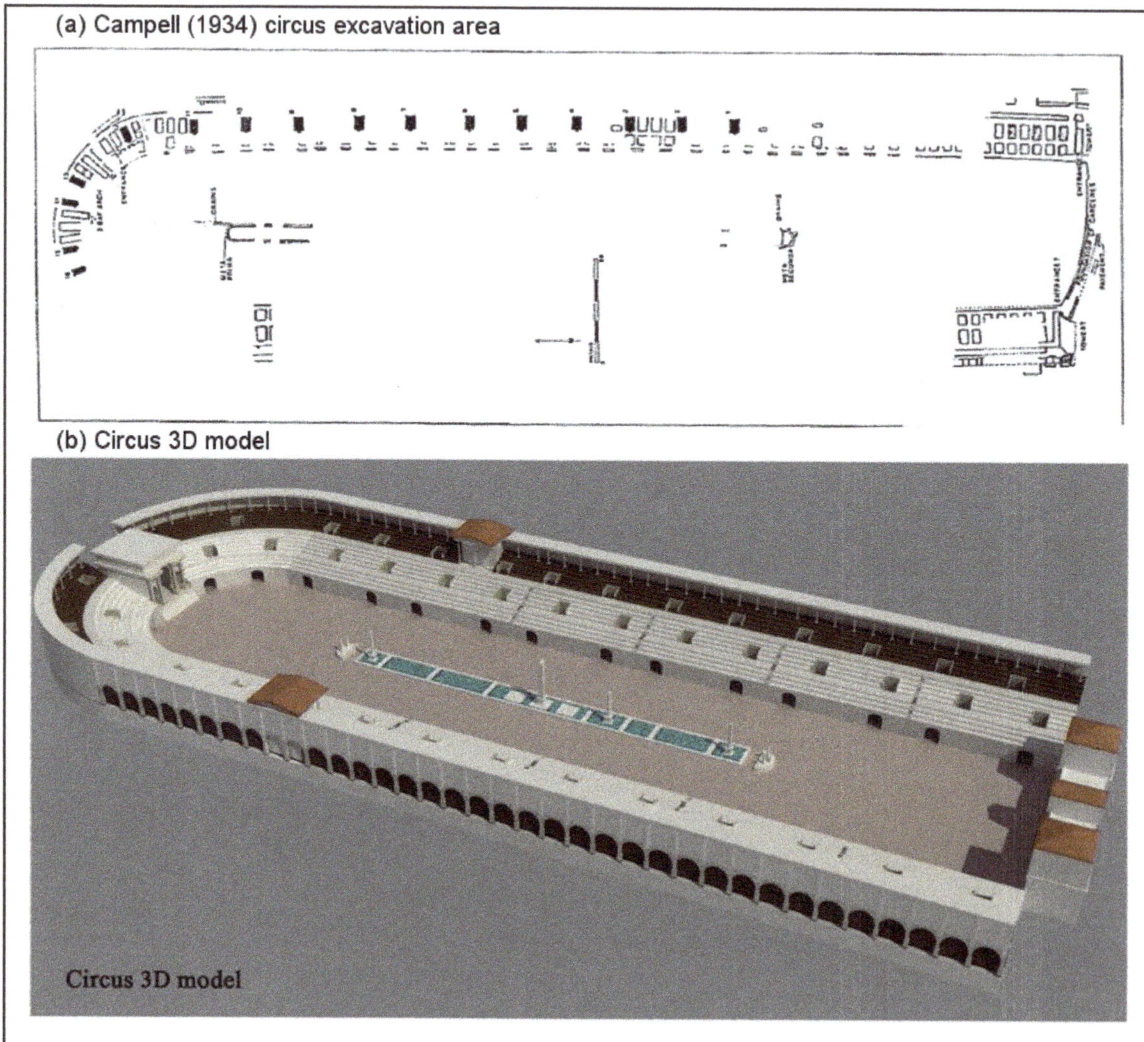

Figure 8. Campell's (1934) [21] circus excavation area and model produced from his plan.

Using the plan, the excavation team discovered the eastern and northern parts of the Hippodrome and the internal and external walls of the arena section. In the excavations, the podium, primum maenianum, ambulacrum, and secundum maenianum were located. Furthermore, these excavations demonstrated that external walls were supported by columns to create a gallery.

3.4. The Modeling of Ancient Antioch's Aqueducts and Nymphaeums

In addition to its great buildings, Ancient Antioch is also famous for its water structures. According to information obtained by Demir (1996) [42], Antioch's need for water was satisfied by the Daphne waterfalls, which flowed 2.8 km from the city. Water was transported through ducts and bridges. The water flow emerged from the channels (ducts and water bridges) and reached important tunnels and regions along the city.

Historical gravures and photographs are used in modeling and dimensioning aqueducts. Downey's (1961) [32] and Uggeri's (1996) [33] maps used in other studies aided in modeling water routes. Topographical structures between Daphne and Antioch were examined carefully, and an attempt was made to recreate these structures authentically (Figure 9).

Figure 9. Ancient Antioch's modeled aqueducts from the Daphne waterfalls.

Water reaching the city through ducts was collected in pools called nymphaeums. Thus, water was used functionally and esthetically; the nymphaeums, varying in size, served as gathering centers for Antioch's population.

3.5. The Modeling of Ancient Antioch's City Walls

Significant parts of Antioch's walls still remain, especially in the east, over the crest of Mount Silpius. Much of the visible remains are dated to the Justinian walls of the sixth century; Justinian reduced the size of the city walls to align them with the much smaller Antioch of his time. However, several parts of what Hellenistic and Roman walls might be are also visible [43].

Antioch's city walls were constructed in the 300s BC during the time of Seleukos I Nikator, who was among the generals of Alexander the Great. However, the walls took their essential structure and current status during the Justinian period. Estimated to be greater than 20 km in length, the city walls accommodated more than 300 towers [44].

Written resources and gravures were mainly assessed for modeling Ancient Antioch's city walls. Gravures drawn in several periods were particularly useful for modeling and dimensioning city walls and towers (Figure 10).

Figure 10. An 18th-century gravure by Louis-Francois Cassas [25] depicting Ancient Antioch's city walls (**a**), and the model produced from this gravure (**b**,**c**).

3.6. Modeling Ancient Antioch's Houses and Baths

The excavation notes and studies from previous researchers (Pamir and Brands, 2005) [45] (Dobbings, 2000) [46] (Kondoleon, 2000) [27] (Gruber and Dobbings, 2013) [28] were mainly used for modeling Antioch's houses and Roman baths. In the excavations of ancient Antioch, the remains of five large baths were found, which were called A, B, C, D, and E. Among them, bath C had the most data about the building (Figure 11).

Figure 11. Model of Roman bath C, Ancient Antioch.

4. Discussion and Conclusions

This research is important for revitalizing lost or destroyed cities, buildings, and their surroundings. Open areas such as roads, pavement, agoras, and avenues—as well as buildings—are modeled. Thus, open urban areas can be perceived in 3D and with depth, enabling experts to produce information and opinions more easily. With character modeling and the creation of dramas, we can reanimate cultural heritage by understanding the dimensional sizes and distances of outdoor social and economic life and recreational areas. Furthermore, this is new work in tracking past cities' historical development because it covers various scientific branches (e.g., archeology, landscaping planning, urban planning, and computer-aided animation). Finally, this is a unique project due to its use of various scientific methods and software integration in fields such as remote perception, solid modeling, and animation.

The greatest limitation throughout the study was obtaining real data about the area or structure to be visualized and modeled. If we had decided to model structures whose dimensions and sizes were certain, only two or three buildings would have been realized, not the whole of Antioch, thus limiting our view of the city. Another problem lay in the absence of important structures' (excavated) remains and their presence beneath several meters of present-day Antakya, where excavations at the desired level are not currently feasible.

Furthermore, the real-time engine (OGRE 3D) used for this study cannot fully operate at the desired level. Despite models produced throughout the study being transferred into the engine, problems occurred while localizing coatings because the software in which models are created and the coordinating software systems in the engine differ. In addition, the OGRE 3D game engine has been abandoned due to difficulties in use, interface complexity, and, most importantly, problems in the integration of models. After the necessary literature study and research, the Unity 3D game engine was used. This game engine is powerful, works across platforms, and is highly user-friendly to both beginners and experts. Unity 3D is a game engine that is advanced and renders real-world graphics without taking much processing power.

The integration with the Unity software, high-resolution panoramic images, and 3D model reconstructions is very successful [47]. Challenges to large-scale animation data visualization, including parallel animation data redundancy, huge data size, dynamic user interaction, and portability, are overcome through algorithms specifically designed for preprocessing simulation data and visualization modules optimized using the powerful 3D rendering and interactive game engine Unity 3D. A comparison of this study with various visualization tools and realistic simulations commonly used to visualize large-scale idealized datasets highlights the visualization efficiency, seamless interaction, and lightweight properties of Ancient Antioch.

Therefore, foundational research such as this is significant for early results that are rich in visual quality. We can expect constantly improving images for the preservation, analysis, and interpretation of cultural heritage—whether for the ultimate purpose of introduction, education, or research.

In particular, this research is important for animating ancient cities. The example used here involved recreating Ancient Antakya's historical past and introducing its culture, particularly through presentations in museums and art galleries. The images and animations from this study are used in the Hatay Archeology Museum (Figure 12).

Figure 12. Three-dimensional Antioch images in Hatay Archeology Museum.

Perhaps the most important output of this study was the use of almost all of the images obtained in the Hatay Archaeology Museum. The visuals and animations of this study are shown on the TV screens in the museum exhibition areas and the museum movie hall. After March 2023, it is planned that the presentation of Ancient Antioch will be opened to visitors on special platforms with Unity 3D, which is at the final stage of preparations.

Practical uses of the methodologies and materials include an animation of Antioch during the Roman period for short films and documentaries (Figure 13). In addition, all the animation, images, and drawings obtained through this study will be transferred into an easily accessible DVD format. Therefore, this study not only introduces Antioch but also can be a resource for undergraduates and postgraduates studying history, the history of art, and archeology departments.

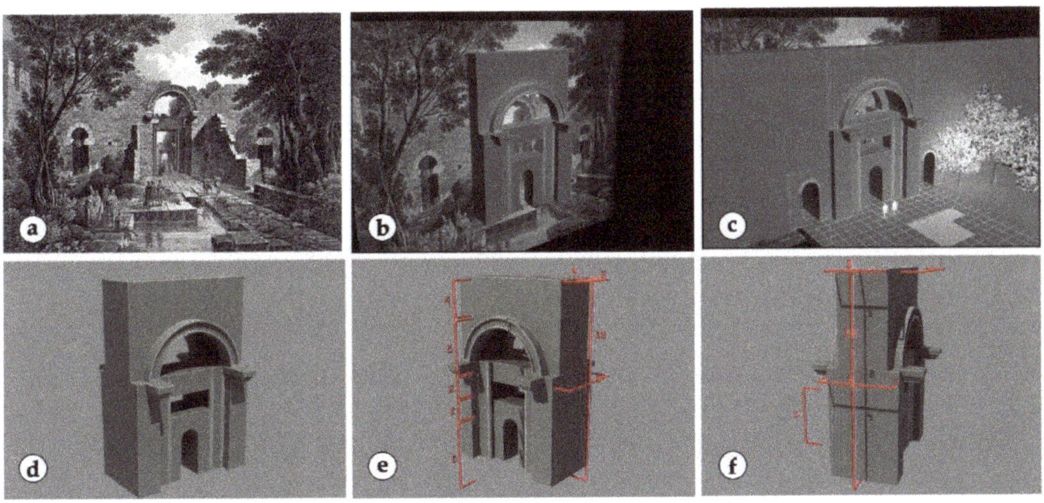

Figure 13. An animation about the Beroea (Medina) Gate: (**a**) Cassas' engraving of the Beroea gate, (**b**,**c**) modeling of the 3d gate from the engraving (**d**–**f**) Animation frames showing the dimensions of the Gate.

Furthermore, the images obtained through this study will be an important data resource for archeology and excavation, providing 3D information and helping determine the direction of excavations, particularly those in progress around Antioch.

This model is not just a visualization either; it helps to reveal new information as well by conjoining previously disconnected knowledge into one place. Our volumetric and multi-perspective view of the ancient city allows us to answer new questions about the tradition, culture, and philosophy of the society, as well as their reciprocal relationship to the built environment. Along with being an academic tool, the model is also a valuable cultural tool that can expand the understanding of the world of Ancient Antioch history and rekindle its place in Rome's Antioch memory and identity.

Finally, this study can serve to motivate similar studies in other regions. For example, when dramas and 3D animations are presented visually during simulation and consultation phases, an excavation team can form opinions and plans about its future work. In other words, a new synthetic process enables an ongoing feedback loop; an actual excavation area's data can be constantly modeled and remodeled, thus improving both the scientific feasibility, quality, and uniqueness of excavation work and the practical dissemination of scientific results.

Funding: Study is funded by 'the Scientific Technological Research Council of Turkey' (TÜBİTAK) as the project numbered 107K278.

Conflicts of Interest: The authors declare no conflict of interest. The funders had no role in the design of the study; in the collection, analyses, or interpretation of data; in the writing of the manuscript, or in the decision to publish the results.

References

1. Ervin, S.M. Digital Landscape Modeling and Visualization: A Research Agenda. *Landsc. Urban Plan.* **2001**, *54*, 49–62. [CrossRef]
2. Koutsoudis, A.; Stavroglou, K.; Pavlidis, G.; Chamzas, C. 3DSSE—A 3D Scene Search Engine. Exploring 3D Scenes Using Keywords. *J. Cult. Herit.* **2012**, *13*, 187–194. [CrossRef]
3. Gutierrez, D.; Seron, F.J.; Magallon, J.A.; Sobreviela, E.J.; Latorre, P. Archaeological and Cultural Heritage: Bringing Life to an Unearthed Muslim Suburb in an Immersive Environment. *J. Cult. Herit.* **2004**, *5*, 63–74. [CrossRef]
4. Bruno, F.; Bruno, S.; de Sensi, G.; Luchi, M.L.; Mancuso, S.; Muzzupappa, M. From 3D Reconstruction to Virtual Reality: A Complete Methodology for Digital Archaeological Exhibition. *J. Cult. Herit.* **2010**, *11*, 42–49. [CrossRef]
5. Rua, H.; Alvito, P. Living the Past: 3D Models, Virtual Reality and Game Engines as Tools for Supporting Archaeology and the Reconstruction of Cultural Heritage—The Case-Study of the Roman Villa of Casal de Freiria. *J. Archaeol. Sci.* **2011**, *38*, 3296–3308. [CrossRef]
6. Fontana, R.; Greco, M.; Materazzi, M.; Pampaloni, E.; Pezzati, L.; Rocchini, C.; Scopigno, R. Three-Dimensional Modelling of Statues: The Minerva of Arezzo. *J. Cult. Herit.* **2002**, *3*, 325–331. [CrossRef]
7. Koutsoudis, A.; Arnaoutoglou, F.; Chamzas, C. On 3D Reconstruction of the Old City of Xanthi. A Minimum Budget Approach to Virtual Touring Based on Photogrammetry. *J. Cult. Herit.* **2007**, *8*, 26–31. [CrossRef]
8. Liritzis, I.; Al-Otaibi, F.M.; Volonakis, P.; Drivaliari, A. Digital Technologies and Trends in Cultural Heritage. *Mediterr. Archaeol. Archaeom.* **2015**, *15*, 313–332. [CrossRef]
9. Besoain, F.; González-Ortega, J.; Gallardo, I. An Evaluation of the Effects of a Virtual Museum on Users’ Attitudes towards Cultural Heritage. *Appl. Sci.* **2022**, *12*, 1341.
10. Schweibenz, W. The Virtual Museum: An Overview of Its Origins, Concepts, and Terminology. Digitisation of Photographs (Art) View Project Personas for Museums View Project. *Mus. Rev.* **2019**, *4*, 1–29.
11. Chrysanthi, A.; Ant, A.; Chrysant, A.; Perry, S.; Roussou, M. Cultural heritage and social experiences in the times of COVID-19. In Proceedings of the 2020 AVI2CH Workshop on Advanced Visual Interfaces and Interactions in Cultural Heritage, Ischia, Italy, 29 September 2020.
12. Marín-Morales, J.; Higuera Trujillo, J.L.; Greco, A.; Guixeres, J.; Llinares, C.; Gentili, C.; Scilingo, E.; Alcañiz Raya, M.; Valenza, G. Real vs. Immersive-Virtual Emotional Experience: Analysis of Psycho-Physiological Patterns in a Free Exploration of an Art Museum. *PLoS ONE* **2019**, *14*, e0223881. [CrossRef]
13. Theodoropoulos, A.; Antoniou, A. VR Games in Cultural Heritage: A Systematic Review of the Emerging Fields of Virtual Reality and Culture Games. *Appl. Sci.* **2022**, *12*, 8476. [CrossRef]
14. Döker, M.F.; Kırlangıçoğlu, C. Promotion of Cultural Heritages Through A Virtual Museum Platform: Case Study Hagia Sophia. *Sak. Univ. J. Sci.* **2018**, *22*, 1337–1345. [CrossRef]
15. Damiano, R.; Lombardo, V.; Lieto, A.; Borra, D. Exploring Cultural Heritage Repositories with Creative Intelligence. The Labyrinth 3D System. *Entertain. Comput.* **2016**, *16*, 41–52. [CrossRef]
16. Dylla, K.A.; Müller, P.; Ulmer, A.; Haegler, S.; Frischer, B. Rome Reborn 2.0 a framework for virtual city reconstruction using procedural modeling techniques. In Proceedings of the 37th Annual International Conference on Computer Applications and Quantitative Methods in Archaeology (CAA), Williamsburg, VA, USA, 22–26 March 2009.
17. Frischer, B.; Guidi, G.; Lucenti, I. Rome Reborn-Virtualizing the ancient imperial Rome. In *Workshop on 3D Virtual Reconstruction and Visualization of Complex Architectures*; Fondazione Bruno Kessler: Trento, Italy, 2007.
18. Denker, A. Rebuilding Palmyra Virtually: Recreation of Its Former Glory in Digital Space. *Virtual Archaeol. Rev.* **2017**, *8*, 20–30. [CrossRef]
19. Denker, A. Vırtual palmyra: 3d reconstructıon of the lost realıty of the brıde of the desert. In Proceedings of the 8th International Congress on Archaeology, Computer Graphics, Cultural Heritage and Innovation 'ARQUEOLÓGICA 2.0', Valencia, Spain, 5–7 September 2016; MIT Press: Cambridge, MA, USA, 2016; pp. 318–321.
20. Bobou, O.; Kristensen, N.B.; McAvoy, S.; Raja, R. Archive Archaeology in Palmyra, Syria a New 3D Reconstruction of the Tomb of Hairan. *Digit. Appl. Archaeol. Cult. Herit.* **2020**, *19*, e00164. [CrossRef]
21. Khorloo, O.; Ulambayar, E.; Altantsetseg, E. Virtual Reconstruction of the Ancient City of Karakorum. *Comput. Animat. Virtual Worlds* **2022**, *33*, e2087. [CrossRef]
22. Ferwati, M.S.; el Menshawy, S. Virtual Reconstruction of the Historic City of Al-Zubarah in Qatar. *Digit. Appl. Archaeol. Cult. Herit.* **2021**, *21*, e00177. [CrossRef]
23. Smith, M.; Walford, N.S.; Jimenez-Bescos, C. Using 3D Modelling and Game Engine Technologies for Interactive Exploration of Cultural Heritage: An Evaluation of Four Game Engines in Relation to Roman Archaeological Heritage. *Digit. Appl. Archaeol. Cult. Herit.* **2019**, *14*, e00113. [CrossRef]
24. Haegler, S.; Müller, P.; van Gool, L. Procedural Modeling for Digital Cultural Heritage. *EURASIP J. Image Video Process.* **2009**, *2009*, 1–11. [CrossRef]

25. Denard, H. *Implementing Best Practice in Cultural Heritage Visualisation: The London Charter BT—Good Practice in Archaeological Diagnostics: Non-Invasive Survey of Complex Archaeological Sites*; Corsi, C., Slapšak, B., Vermeulen, F., Eds.; Springer International Publishing: Cham, Switzerland, 2013; pp. 255–268. ISBN 978-3-319-01784-6.
26. Denard, H. The London charter for the computer-based visualisation of cultural heritage. In *Paradata and Transparency in Virtual Heritage*; Routledge: London, UK, 2009; pp. 99–104.
27. Kondoleon, C. (Ed.) *Antioch: The Lost Ancient City*; Princeton University Press: Princeton, NJ, USA, 2000; ISBN 978-0691049335.
28. Kloeg, S.P. Antioch the Great Population and Economy of Second-Century Antioch. Master's Thesis, Leiden University, Leiden, The Netherlands, 2013.
29. Sbeinati, M.R.; Darawcheh, R.; Mouty, M. The Historical Earthquakes of Syria: An Analysis of Large and Moderate Earthquakes from 1365 B.C. to 1900 A.D. *Ann. Geophys.* **2005**, *48*, 347–435. [CrossRef]
30. Sandwell, I.; Huskinson, J. *Culture and Society in Late Roman Antioch*; Oxbow Books: Oxford, UK, 2003; ISBN 978-1842171028.
31. Kerlow, I. *The Art of 3D Computer Animation and Effects*, 3rd ed.; John Wiley Sons, Inc.: Hoboken, NJ, USA, 2004; ISBN 978-0470084908.
32. Downey, G. *A History of Antioch in Syria: From Seleucus to the Arab Conquest*; Reprint of; Princeton University Press: Princeton, NJ, USA, 1961; ISBN 978-1258486655.
33. Uggeri, G. Antiochia Sull'Oronte Nel IV Secolo d. C. In Proceedings of the Atti II Congress Topografia Antica, Roman Town-Planning, Rome, 15–16 May 1996; pp. 179–222.
34. Lassus, J. *Antioch on-the-Orontes V, Les Portiques d'Antioch. Committee for the Excavation of Antioch and Its Vicinity*; Princeton University Press: Princeton, NJ, USA, 1972.
35. Norman, A.F. *Libanius: Autobiography and Selected Letters*; Cambridge University Press: London, UK, 1993; ISBN 0-674-99527-9.
36. Robertson, D.S. *Greek and Roman Architecture*; Cambridge University Press: London, UK, 1992; ISBN 978-0521094528.
37. Ward-Perkins, J.B. *Roman Imperial Architecture*; Yale University Press: New Haven, Connecticut, 1992; ISBN 978-0300052923.
38. Guidetti, F. Urban continuity and change in late Roman Antioch. In urban decline. In the byzantine realm. In Proceedings of the International Conference Acta Byzantina Fennica, Helsinki, Finland, 25 September 2010; Volume 3, pp. 81–104.
39. Ecclestone, C. Antiochepedia = Musings upon Ancient Antioch. Available online: http://libaniusredux.blogspot.com.tr/html. (accessed on 15 October 2014).
40. Campbell, W.A. Excavations at Antioch-on-the-Orontes. *Am. J. Archaeol.* **1934**, *38*, 201–206. [CrossRef]
41. Wilmott, T. *Roman Amphitheatres and Spectacula, a 21st-Century Perspective: Papers from an International Conference Held at Chester, 16–18 February 2007*; Archaeopress: Oxford, UK, 2009; ISBN 9781407304267.
42. Demir, A. *Çağlar Içinde Antakya*; Akbank Kültür ve Sanat Kitapları: İstanbul, Turkey, 1996.
43. Brands, G. *Antike Welt*; Zeitschrift für Archäologie und Kunstgeschichte: Darmstadt, Germany, 2004; Volume 35, pp. 10–16.
44. Williams, S. *Diocletian and the Roman Recovery*; Routledge: London, UK, 2000; ISBN 0415918278.
45. Pamir, H.; Gunner, B. The Asi delta and the Asi valley archaeological project in 2004: Samandağ and antakya surveys. In Proceedings of the News of Archaeology from Anatolıa's Medıterranean Areas (ANMED), Ankara, Turkey, 29–30 May 2005; pp. 103–108.
46. Dobbins, J. The houses of Antioch. In *Antioch: The Lost Ancient City*; Kondoleon, C., Ed.; Princeton University Press: Princeton, NJ, USA, 2000; pp. 74–82. ISBN 978-0691049335.
47. Clini, P.; Ruggeri, L.; Angeloni, R.; Sasso, M. Interactive immersive virtualmuseum: Digital documentation for virtual interaction. In Proceedings of the International Archives of the Photogrammetry, Remote Sensing and Spatial Information Sciences—ISPRS Archives, Riva del Garda, Italy, 30 May 2018; Volume 42, pp. 251–257.

applied sciences

MDPI

Article

A Methodical Approach to 3D Scanning of Heritage Objects Being under Continuous Display

Jacek Kęsik [1], Kamil Żyła [1], Jerzy Montusiewicz [1], Marek Miłosz [1,*], Calin Neamtu [2] and Marta Juszczyk [3]

[1] Department of Computer Science, Faculty of Electrical Engineering and Computer Science, Lublin University of Technology, Nadbystrzycka 36B, 20-618 Lublin, Poland
[2] Department of Design Engineering and Robotics, Technical University of Cluj-Napoca, 400114 Cluj-Napoca, Romania
[3] Department of Information Systems Engineering, Faculty of Management, Lublin University of Technology, Nadbystrzycka 38, 20-618 Lublin, Poland
* Correspondence: m.milosz@pollub.pl

Abstract: Three dimensional digitization of cultural heritage resources gains a lot of attention from the European Union and the United Nations, which is clearly revealed in current strategic goals and financing perspectives. Existing methodological approaches to 3D scanning in a prevailing number of cases assume that the procedure of scanning is performed in places that are closed to tourists, at least for the time of scanning. However, closing an exhibition for tourists or moving an artifact to be scanned is not always possible. Thanks to the long-term experience of the authors with 3D scanning of cultural heritage, the special procedure was designed for small and medium size objects to overcome difficulties expected in such cases. The procedure has been successfully implemented during 3D scanning of objects exhibited in the Silk Road region (on the territory of modern Uzbekistan), as well as objects being parts of wooden sacral architecture of the Maramures region (in Romania). It was revealed that the proposed procedure was successfully allowed to counteract organizational problems during 3D scanning of heritage objects being under continuous display, and that the achieved results of scanning were nevertheless of good quality.

Keywords: 3D scanning; cultural heritage; methodical approach; exhibitions under continuous display; Silk Road; Maramures wooden architecture

Citation: Kęsik, J.; Żyła, K.; Montusiewicz, J.; Miłosz, M.; Neamtu, C.; Juszczyk, M. A Methodical Approach to 3D Scanning of Heritage Objects Being under Continuous Display. *Appl. Sci.* **2023**, *13*, 441. https://doi.org/10.3390/app13010441

Academic Editor: Ion Sandu

Received: 3 December 2022
Revised: 18 December 2022
Accepted: 26 December 2022
Published: 29 December 2022

1. Introduction

1.1. Digitization of Cultural Heritage

Three dimensional digitization of cultural heritage resources is gaining a lot of attention from international and governmental institutions, which is clearly revealed in their current strategic goals and financing perspectives. The European Union is one of the entities very active in this field, to mention the initiative called "New European BauHaus" aimed at building a sustainable and inclusive future among others by addressing complex societal problems and creating a bridge between science, technology and culture [1]. Maintaining heritage within the shared European cultural space and adapting it for the future is a key concept [2]. Funding for heritage preservation is granted under the Horizon Europe as well, for example the "Culture, creativity and inclusive society" programme [3]. Another example is the United Nations and its Development Programme (Standard 4: Cultural Heritage). Cultural heritage preservation, protection and promotion in a manner consistent with the UNESCO cultural heritage conventions is mentioned as the aim [4].

1.2. The Problem of 3D Digitization of Historical Objects in Museums

Existing methodological approaches to 3D scanning in a prevailing number of cases assume that the procedure of scanning is performed in places that are closed to tourists, at least for the time of scanning. Such an approach gives the best results, thanks to the time available and the possibility of free positioning of a scanner. Such conclusions might be derived from the authors' own works [5–11] as well as from others [12–15].

However, closing an exhibition for tourists, or moving an artifact to be scanned, is not always possible. Such an action might be unacceptable for a museum. Additionally moving an artifact might not be possible, e.g., because it is part of an architectural object, a collection to which the artifact belongs is in permanent exhibition, moving the artifact is too difficult or risky, a procedure of loaning it creates too much trouble and risk, etc. In such cases, the only thing left to consider is scanning at the point of exposure of the objects (Figure 1).

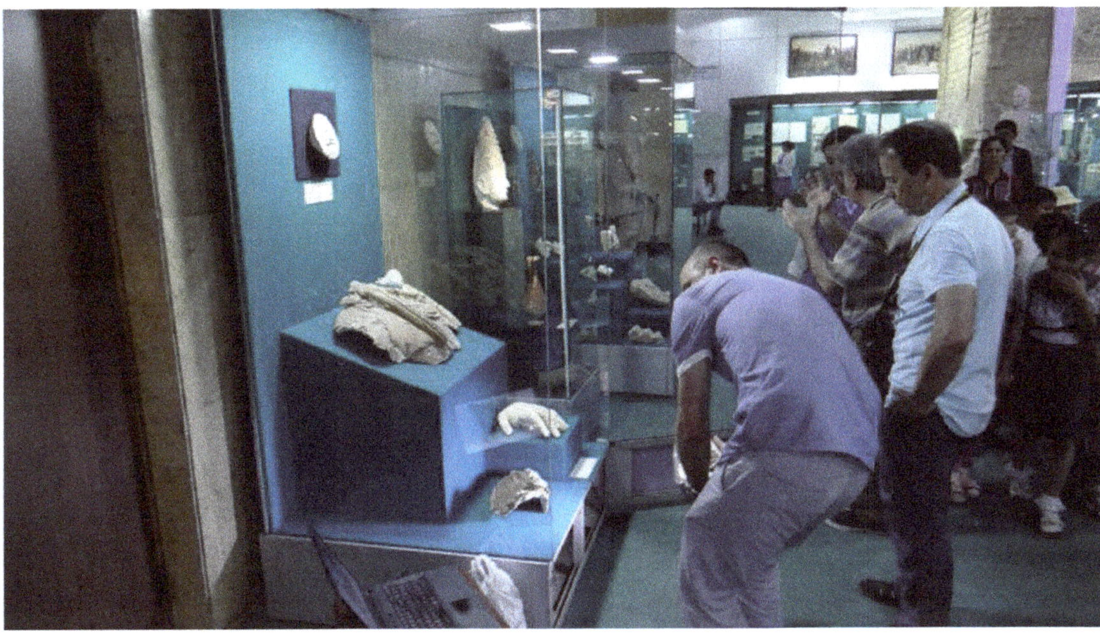

Figure 1. 3D scanning under tourist traffic in Tashkent at the National Museum of the History of Uzbekistan.

1.3. Study Motivation

The authors have experienced the above-mentioned problems during their work and successfully found appropriate solutions. As a result, they have developed a methodical approach (called "3DScaMITE"—3D Scanning Methodology in Intensive Tourism Environment) for scanning large architectural objects under continuous tourist traffic and applied it successfully in scanning the Timurid architecture in Central Asia [16]. Nevertheless, the gap concerning a methodical approach to 3D scanning of small and medium sized objects under continuous display is still present.

Thus, the authors decided to present in this work the methodical approach to 3D scanning of small- and medium-sized heritage objects in the cases when closing an exhibition for tourists, or moving an object to be scanned, is not possible. The methodology was created in order to overcome difficulties expected in such cases, as a result of long-term experience of the authors with 3D scanning of cultural heritage objects. It was developed and verified on the basis of experimental implementation of 3D scanning sessions in state museums of Uzbekistan (Afrasiab in Samarkand and in Tashkent) as well as during the scanning of wooden sacral architecture in the Maramures region in Romania. Additional value of this work, aside from introducing a new methodology, is complying with the goals and needs of European Union and United Nations initiatives.

2. Study Background

In situ digitization of heritage objects is usually affected by the following difficulties: the necessity to bring and set-up the equipment needed during digitization; limited access to heritage objects and safety of their extraction; as well as ongoing tourist traffic.

Digitization of tangible cultural heritage objects can be implemented in many ways, in terms of technology as well as conditions for data acquisition and access to heritage site facilities. Daneshmand et al. in [17] present a variety of technological solutions suitable for 3D digitization. Most of the solutions are based on aerial and terrestrial photogrammetry, although the role of terrestrial laser scanning usage in data acquisition is mentioned as well. The authors also give examples of 3D reconstruction of architectural structures, i.e., the remains of Las Higueras city (Argentina) and historical buildings such as the Rohan Palace facade, the Catholic church of St-Pierre-le-Jeune, the Josephine Pavilion in Strasbourg (France), the Byzantine Land Walls of Istanbul (Turkey) or the Temple of the Sacred Tooth Relic at Kandy (Sri Lanka). Architectural details were acquired by means of a hand-held scanner using Artec's structured light technology. However, the authors did not pay attention to the aspects related to local (in situ) conditions concerning (influencing) the digitization process, despite the fact that these objects were frequently visited by tourists.

Miłosz et al. in [18] present various theoretical aspects regarding the issues of 3D digitization of architectural and museum objects, as well as many practical implementations. Some of the described devices have significant limitations in their use in situ, e.g., stationary chamber scanners (Roland PICZA 3D Laser Scanner LPX-600) have a small working space and cause difficulties in transport due to their large weight and size. Another presented device—ZScanner ® 700, which is a hand-held laser scanner, requires placing markers on the surface of a digitized object, which is usually not allowed in the case of historic objects [19]. A confirmed drawback of these devices is the inability to collect information about the texture of scanned objects, which limits their use in transferring cultural heritage to the digital world. Kowalski et al. in [20] present a low-cost 3D data acquisition system for small objects using a set of Kinect v2 devices cooperating with the open source LiveScan3D software. Unfortunately, there are a lot of cases requiring more professional solutions in terms of precision, range, durability, ability to deal with environmental factors.

Tucci et al. in [21], apart from the above-mentioned aspects, deal with the topic of 3D reconstruction. Practical implementation of activities concerning the 3D digitization of heritage objects can be found in many other works, to mention creating virtual spaces [22,23] or the scanning of: the Hercules statue from the Antalya Museum in Turkey and the Khmer head from the Rietberg Museum in Zurich [24]; a complex of historic buildings in Malacca, Malaysia, declared a UNESCO World Heritage Site in 2008 [25]; artifacts in the area of Sarmisegetuza Regia archaeological site [26]. Nevertheless, in the opinion of the authors of this article, the topic of 3D scanning of heritage objects being under continuous display was not addressed as it needs to be. What is more, there is a wide agreement that there are not enough free databases that could be used for comparative tests of algorithms and procedures [17,27–29].

Few works address issues related to accessing heritage objects and dealing with particular conditions of heritage sites subjected to 3D scanning. Montusiewicz et al. in [5] present many problems related to removing objects to be scanned from an exhibition and preparing them for 3D scanning process. In order to extract the objects from the showcases from the 1970s, several museum employees offered their cooperation. Some objects needed re-forming after being extracted. Construction of the scanning stage required a lot of ingenuity and depended on the nature of the object. On the other hand, there is the case of the County Museum in Nysa [30], where wooden polychrome sculptures from the 16th/17th centuries were digitized. The work was carried out in exhibition halls with diffused lighting, where the objects were put on special pedestals without additional covers, with omnidirectional access. Due to the prevailing COVID-19 pandemic, there were no visitors in the museum halls, which made it much easier to carry out all the works.

Another kind of problem, when 3D digitizing heritage objects is ongoing tourist traffic, resulting in visitors being scanned altogether with the object [16,31,32]. Girelli et al. in [31] the describe usage of photogrammetry to generate a detailed photo-textured 3D model of a fountain. The result, which was noisy due to the constant presence of people, was cleaned by applying a manual procedure. Miłosz et al. in [16] present the problems of digitizing large architectural objects in Registan (Samarkand, Uzbekistan), using Faro's terrestrial laser scanner, with intensive tourist traffic. The method of removing visitors from scans is described by Kęsik et al. in [32].

Summing up, the information obtained from the analyzed articles, it appears that the issue of 3D scanning in in situ conditions is treated marginally. This may be the result of the fact that the 3D digitization of tangible cultural heritage is incidental, not systematic. Preparing operation procedures for teams undertaking these activities and describing the various emerging problems is an important issue when undertaking continuous digitization by many independent teams in various museums around the world. Another aspect which should be addressed in the future is the sustainability aspects concerning heritage sites and their surroundings [33–37].

3. Research Question

Persons working in the real environment on 3D digitization of heritage artifacts being under continuous display encounter a common set of problems. According to the authors' experience these might be:

- work in the conditions of ongoing tourist traffic,
- logistical and legal problems related to the removal of exhibits from the exhibition hall,
- the need to organize temporal scan locations,
- inability to move some exhibits,
- inability to remove some exhibits from the showcase.

Some of the abovementioned problems were addressed in the case of large architectural objects by the "3DScaMITE" methodology [16]. However, there is a still need to address these problems in the case of small- and medium-sized objects. In order to fill the identified gap, the authors in this work propose a new methodical approach (called "3DScaMOTO") and pose the following research question:

Is it possible to acquire, in an effective way, 3D scans using 3D scanners, specifically based on structured light, of historical objects while not removing these objects from their original area of exhibition?

In order to answer the research question and verify the proposed method, the authors conducted a number of 3D scanning sessions of the Central Asia and Maramures regions heritage. From the huge variety of artifacts, the following were chosen as the case studies: the figure of two snakes from the State Museum of the History of Uzbekistan in Tashkent, the statue of Buddha from the Fayaztepa complex (Uzbekistan) and the icon of Saint Nicholas of Myra in Păușa (Romania).

4. Materials and Methods

4.1. Identification of Object Classes in Terms of Posed Scanning Difficulties

Based on the authors' experience, the heritage sites may contain several types of objects in terms of the possibility of scanning them. To name them:

- A. It can be moved from an exhibition place to an ad hoc scanning studio. Objects of small dimensions and weight, in condition (shape) that allows them to be moved within the exposure area in accordance with the procedures indicated by a heritage site.
- B. It cannot be moved from an exhibition place. It has to be scanned at the place of exposure, but there is free access to it. The object cannot be moved, for example, because of: its size and weight, its permanent connection with the rest of the exhibition (e.g., base-relief), no permission to move the object due to its condition or even an alarm system.

- C. It cannot be moved from the exhibition place. It has to be scanned at the place of exposure and access to it is difficult. Such objects are B-type objects, although the access to them is additionally hampered by non-removable physical protections in the form of display cases, windows, bars, etc.
- D. It cannot be scanned at all. No access and/or consent from the facility is granted.

The "3DScaMOTO" methodology, presented in detail in the next subsection, takes into account the presence of all the three types of objects (A, B, C) for which 3D scanning is potentially possible.

4.2. Description of the "3DScaMOTO" Methodology

This section introduces a methodical approach for 3D scanning of small- and medium-sized objects under continuous display. It was called "3DScaMOTO", which means "**3D Sca**nning **M**ethodology in the **O**pen for **T**ourists **O**bjects". See Table 1 for details on particular stages and their outcomes. Groups A, B, C from the table refer to object types characterized in Section 4.1. The methodology is not limited exclusively to museums but is valid for all heritage sites that are open for tourist traffic and for objects of dimensions that are within the capabilities of structured-light 3D scanners.

The initial stage (number 1—Planning) of the methodology covers the preliminary steps before actual 3D scanning. The cultural heritage site (which can be a museum, a historical place or any other site where heritage objects are placed) has to be examined and objects to be scanned have to be chosen. Once the list of objects is created the team of specialists, together with representatives of the site authorities, assign the objects to the A, B, C, and D groups.

In the case when some objects are assigned to the A group, the place for the future ad hoc scanning studio is chosen. The requirements for media (e.g. availability of electric current) are agreed according to each group of objects. At this stage the list of chosen objects is reconsidered taking into account the media limitations, constraints imposed by the site authorities and the feasibility assessment of the scanning. The result is the final, approved list of objects. The list is then rearranged according to the planned order of scanning and the naming convention is chosen so that thematic groups can be easily distinguished.

Onsite replanning takes place just before the scanning session. The agreed conditions are checked, and countermeasures are implemented if any problems arise. A verified scanning sequence plan is created, taking into account any changes in the sequence forced by the situation (e.g., the need to temporarily restore tourist traffic). After that, the 2nd stage can be initiated.

The 2nd stage is entirely about the proper 3D scanning execution. It is the last stage that has to be carried out at a heritage site—further stages might be executed in convenient time and place. If there are any type A objects on the list, the ad hoc studio is set-up to provide the best possible environment for scanning. The 3D scanning procedure is executed slightly differently, depending on the type of object being scanned. For type A objects, 3D scanning is carried out in the following sequence: transferring the object to the scanning station in the ad hoc studio (carried out by heritage site staff), selection of parameters, actual scanning, initial check of the results (if there are problems, scanning of the relevant fragments is repeated), collection of additional data about the scanned object (photographic and descriptive documentation), returning the object to the exhibition site (again carried out by heritage site staff).

In the case of type B objects, the 3D scanning procedure begins with preparation of the exhibition site (e.g. removing obstacles and setting additional lighting). The scope of the exhibition site modification depends on the decision of heritage site staff. Permission has to be obtained each time for each modification. Then, the object is 3D scanned, taking into account the constraints of a given place. The rest of the procedure is analogous as in the case of type A objects. Results should undergo preliminary checking—problematic places might be scanned again if necessary. When the scanning process is completed, the object's exhibition site has to be restored to its normal condition.

Table 1. 3DScaMOTO—methodical approach description.

Stage Name	Substages	Results
1. Planning	1.1. Strategic planning 1.1.1. Getting to know the site, objects and exhibitions 1.1.2. Selecting objects to scan 1.1.3. Acquiring information about selected objects 1.1.4. Dividing objects into groups A, B, C 1.1.5. Determining the order of scanning	List of objects to be scanned divided into groups (A, B, C) and a preliminary sequence of scanning
	1.2. Initial planning 1.2.1. Selecting places for ad hoc studios 1.2.2. Determining conditions of B and C type object sites 1.2.3. Assessing feasibility of the scanning—rejection of the impossible 1.2.4. Redetermining and confirming the order of scanning 1.2.5. Setting naming conventions	Final list of objects Initial order of scanning
	1.3. On-site replanning 1.3.1. Identifying expected problems 1.3.2. Rearranging plan of the scanning 1.3.3. Marking problematic objects	Redefined order of scanning adjusted to the current on-site conditions
2. Scanning	2.1. Setting-up an ad hoc scanning studio at an exhibition site, if type A objects present	Scanning place for objects of type A
	2.2. For each object of type A: 2.2.1. Moving an object to an ad hoc studio 2.2.2. Setting parameters of scanning 2.2.3. Performing a 3D scan 2.2.4. Checking the 3D scan quality 2.2.5. Marking the 3D scan as problematic (if any, go to 2.2.2) 2.2.6. Acquiring additional data 2.2.7. Returning the object to its original place	Files with scans data Additional data (e.g., photographs, historical background, etc.)
	2.3. For each object of type B: 2.3.1. Preparing an exhibition place of an object for scanning 2.3.2. Setting parameters of scanning 2.3.3. Performing a 3D scan 2.3.4. Checking the 3D scan quality 2.3.5. Marking the 3D scan as problematic (if any, go to 2.3.2) 2.3.6. Acquiring additional data 2.3.7. Restoring the object's exhibition site to its normal condition	Files with scans data Additional data (e.g., photographs, historical background, etc.)
	2.4. For each object of type C: 2.4.1. Preparing an exhibition place of an object for scanning 2.4.2. Setting parameters of scanning 2.4.3. Performing a 3D scan 2.4.4. Checking the 3D scan quality and inconsistencies caused by limited access 2.4.5. Developing an alternative way of problematic areas scanning if necessary (if any, go to 2.4.2) 2.4.6. Acquiring additional data 2.4.7. Restoring the object's exhibition site to its normal condition	Files with scans data Additional data (e.g., photographs, historical background, etc.)

Table 1. *Cont.*

Stage Name	Substages	Results
3. Processing data	3.1. Preparing scans 3.1.1. Analyzing gathered data 3.1.2. Discarding unusable or failed scans 3.1.3. Confirming correctness of an object name and description	Scan data ready for processing
	3.2. Full processing 3.2.1. Filtering and cleaning up data 3.2.2. Curing or removing problematic areas of partial scans 3.2.3. Aligning partial scans 3.2.4. Global-registering of point clouds	Final point cloud
	3.3. Final mesh processing 3.3.1. Setting a processing region 3.3.2. Performing conversion 3.3.3. Cleaning-up and fixing of a resulting 3D model 3.3.4. Coloring of a 3D model surface	Base (high quality) 3D mesh model
4. Preparing for dissemination	4.1. Preparing dissemination 3D model 4.1.1. Obtaining requirements for a dissemination 3D model 4.1.2. Converting a base 3D model to a dissemination model 4.1.3. Testing quality (go to 4.2 if needed) 4.1.4. Arranging a dissemination model in a viewing environment 4.1.5. Preparing/Finalizing a 3D model for distribution	Dissemination 3D model

In the case of type C objects, the procedure is analogous to the case of type B objects. The difference is taking into account difficulties in accessing an object to be scanned. Only during the scan, it is possible to determine how much the existing limitations interfere with the proper scanning process. Some counteracting methods could be introduced depending on the situation, e.g., scanning from a different position, making multiple scans of difficult area fragments, using a different scanning device and/or additional handles.

Stage 3 is about transforming gathered 3D scans into a high-resolution 3D mesh model of the scanned object (called 3D base model). During the scan preparation, the collected data for individual scanned objects are analyzed. Correctness of the object tagging (name and description) is confirmed, and potential errors in the description detected. 3D scans are analyzed in terms of their suitability for building a base 3D model. Rejecting inaccurate or low-quality scans saves time during actual processing. The result of this step are verified object scan data, ready for processing.

Another step is actual processing of scan data (full processing). The data are first cleared of unnecessary elements, then partial scans are aligned to each other. After successful alignment, a global point cloud of the entire object is generated.

During the final mesh processing, the point cloud is converted to a mesh of triangles. Obtained 3D model is cleared of unnecessary elements generated by the triangularization algorithm. It might also be analyzed against mesh discontinuities that could later undergo a fixing procedure. At last, color and texture are applied in order to make the 3D model look like the original object.

Stage 4 can be performed for 3D models produced in stage 3. It is not mandatory—it is executed as the need arises to present (disseminate) obtained results. First, the required technical parameters of the presentation are analyzed. In the next step, the base 3D model is converted to its dissemination version. The result has to be tested to see if it meets the requirements of presentation. If not, it has to be remade. Sometimes it has to be repeated multiple times. When the achieved effect is satisfactory, the model can be arranged in the presentation area and published. Very often dissemination models are of worse quality than base models, in order to comply with performance limitations of technologies used to create the presentation.

4.3. Case Studies Description

The 3DScaMOTO methodology can be used in 3D digitization independently of the way objects are exhibited at a heritage site. The authors have tested the method during implementation of 3D scanning activities concerning the historical area of the Silk Road and wooden sacral architecture of Maramures region in Romania. This section presents three case studies (one for each heritage object type), when the method was applied for type A, B, and C objects being under continuous display.

4.3.1. Case Study 1—"Snake"

Case study 1 concerns the heritage object called "Snake", which belongs to the collection exhibited by the State Museum of the History of Uzbekistan in Tashkent. This is a type A object of the highest level of accessibility. It was scanned using the 3DScaMOTO method during one of the scientific expeditions of the Lublin University of Technology to Central Asia.

"Snake" is a figure depicting two snakes (Figure 2). It is made of polished stone, and its size is 26.3 × 23.6 × 4.4 cm. Due to the relatively small size and weight as well as quite compact construction, the museum authorities agreed to temporarily move the object to an ad hoc scanning site (Figure 3). As a result, the object could be placed on a soft foam in an area of good lighting. The procedure of moving and returning the object was carried out by employees of the museum and each time took about 10 min. The scanning process lasted 20 min, during which nine partial scans of both sides of the object were made. Data processing leading to the high-quality textured 3D mesh model took 2.5 h. The dissemination model, satisfactory in terms of quality and performance ratio, was obtained after 1 h.

Figure 2. "Snake" figure in the State Museum of the History of Uzbekistan in Tashkent.

Figure 3. Scanning environment arrangement to scan the "Snake" figure.

4.3.2. Case Study 2—"Icon of Saint Nicholas of Myra"

The icon of Saint Nicholas of Myra (Figure 4) is an example of the freely accessible but unmovable item. It is fixed to the wall of the orthodox church in Păușa, Romania. This wooden church was built in 1730 on a rectangular plan and restored in 1966–1968. The colorful interior paintings were made in 1800, probably by Ioan Pop from Ungurasi. Saint Nicholas, also known as Nicholas of Bari, is a saint of both the Catholic and Orthodox churches, who lived at the turn of the 3rd and 4th century and was the bishop of Myra. The icon, probably dating from the mid-18th century, is inside a wooden frame with the motif of twisted columns. A characteristic element is a wooden convex halo and gilded carved crosses arranged in diagonal coffers [38]. It was scanned using the 3DScaMOTO method during The First Scientific Expedition of the Lublin University of Technology to the Maramures region.

Due to the missing information regarding the icon's vulnerability to strong light, the natural lighting of the church was utilized while scanning. As a result, exposure to excessive light produced by photographic lamps was avoided. The scanning process lasted 12 min, during which four partial scans of the icon's front side were made. The icon was fixed to the wall; thus its back side was not scanned. It was considered then as a non-mandatory activity. The scanning site did not have to be restored to the original shape, and nothing was moved from the original place. Data processing leading to the high-quality textured 3D mesh model took 1.5 h. The dissemination model, satisfactory in terms of quality and performance ratio, was obtained after 1 h.

4.3.3. Case Study 3—"Buddha"

Case study 3 concerns the heritage object called "Buddha", which belongs to the collection exhibited by the State Museum of the History of Uzbekistan in Tashkent. This is a type C object of low level of accessibility. It was scanned using the 3DScaMOTO method during one of the scientific expeditions of the Lublin University of Technology to Central Asia.

The object is a statue (sculpture) depicting Buddha, sitting in a pose of peace and spiritual harmony under the sacred Bodhi tree, and two monks standing on either side (Figure 5). The statue size is 74.6 × 65.9 × 29.6 cm. It is a most valuable finding from the Fayaztepa complex (Uzbekistan) that was discovered in 1963. The sculpture, carved in limestone and gilded, dates from the 1st century AD. Today, this sculptural composition is one of the most valuable exhibits of the State Museum of the History of Uzbekistan in Tashkent.

Figure 4. The icon of Saint Nicholas of Myra in the orthodox church in Păușa, Romania.

Figure 5. "Buddha" statue in the State Museum of the History of Uzbekistan in Tashkent.

The statue is standing on a pedestal in a display case with glass walls. Access to the statue is possible by opening the side wall on the right side (Figure 6). The statue is too heavy to be easily removed from the display case. During the planning stage, it was estimated that the front glass of the display case was located at a distance from the statue's surface allowing for a 3D scanner to be placed inside. There was also free access to the right side of the statue after opening the display case. The panes on the left side and on the back of the statue were too close to the object for a scanner to be placed inside. The Artec Eva scanner was able to scan the statue's front with satisfactory accuracy, but there was a risk of not being able to scan deep, narrow grooves and internal fragments of the statue due to the dimensions and working distance of the scanner. Additionally, scanning through the glass might have introduced some errors, despite being possible for structured light scanners.

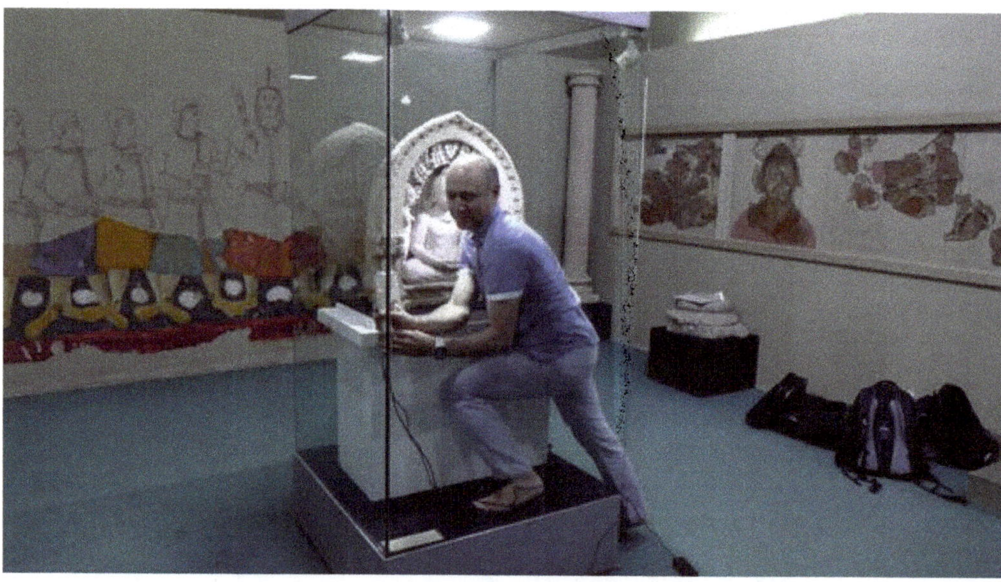

Figure 6. Scanning environment arrangement for scan the "Buddha" statue.

Finally, after consultations with the museum, it was decided to 3D scan the statue's front from the inside of the display case in order to ensure good-quality archiving of the shape of the statue's most valuable parts. In the case of other parts, as of quite simple shape and texture, they were scanned through the glass. Then, the scanning stage was executed. Preparation of the scanning site, including opening the display case, took 10 min. The scanning process itself took 45 min. As a result, 20 partial scans were collected in total: 15 scans of the statue's front carried out directly and 5 scans carried out through the glass. After another 15 min, the display case was restored to its original condition. Data processing leading to the high-quality textured 3D mesh model took 5.3 h. The dissemination model, satisfactory in terms of quality and performance ratio, was obtained after 1.2 h.

4.4. Hardware and Software

The scanning was carried out using a hand-held structured-light Artec EVA 3D scanner, providing 50–100 microns accuracy and saving textures. Photographic documentation was made using a Nikon D7200 digital camera. In the case of several scans of the "Snake" figure, a rotary table with a surface texture was used to facilitate the positioning of the scanner.

The software used to obtain both base and dissemination models was Artec Studio Professional, Blender and MeshLab. Artec Studio was especially important during the scanning and the first phases of scans processing, as a software dedicated for the Artec EVA

3D scanner. The remaining programs were used to process 3D models and export them in proper quality.

Computations were performed on a laptop equipped with i9-9900K processor, 64 GB RAM, RTX 2080 8GB graphics card and 2TB SSD. Using a less powerful computer is still possible, although the time of computations strongly depends on the CPU and availability of sufficient memory to keep the object's data loaded. Data gathering could also be assisted by a MicroSoft Surface tablet—the main advantages are improving portability and easing operation of the whole setup.

5. Results and Discussion

The results of 3D scanning, according to the presented procedure, aggregated for all case studies, are presented in Table 2. The size of a 3D model is provided in MB and thousands of triangles—K stands for thousands and Tri stands for triangles. The processing time is the sum of obtaining 3D base model time and obtaining 3D dissemination model time. Preparation time concerns the time needed to prepare a heritage object for scanning. Restoring time concerns the time needed to restore a heritage site to the condition before preparation for scanning. Discussion on procedure, problems and solutions, in the context of each case study, is presented in the subsequent sections.

Table 2. Results of 3D scanning for the case studies.

Case Study Number	**1**	**2**	**3**
Object name	"Snake"	"Icon of Saint Nicholas of Myra"	"Buddha"
Place of scanning	Tashkent, Uzbekistan	Romania	Tashkent, Uzbekistan
Scanning environment	Ad hoc studio	On the site	On the site
3D scanner used	Artec Eva	Artec Eva	Artec Eva
Scanned object type	A	B	C
Scanned object dimensions	26.3 × 23.6 × 4.4 cm	87.1 × 64.9 × 3.7 cm	74.6 × 65.9 × 29.6 cm
Partial scans number	9	4	20
Rejected scans number	1	0	2
Raw scan total size	0.98 GB	0.38 GB	5.28 GB
Scan average accuracy	0.5 mm	0.6 mm	0.4 mm
Problems with obtaining color and texture	NO	NO	NO
Object surface digitized	100%	~50%	85%
3D base model size	27.2 MB, 394.5 KTri	150.5 MB, 1.03 MTri	62.1 MB, 886.8 KTri
3D dissemination model size	10.6 MB, 145.9 KTri	25.1 MB 250 KTri	10.9 MB, 150.7 KTri
Preparing time	10 min	10 min	10 min
Scanning time	20 min	12 min	45 min
Restoring time	10 min	-	15 min
Processing time	3.5 h	1.5 h	6.5 h
Obtaining 3D base model time	2.5 h	1.0 h	5.3 h
Obtaining 3D dissemination model time	1.0 h	0.5 h	1.2 h

5.1. Case Study 1—"Snake"

In the case of the "Snake" figure, the scanning time was 20 min. During this time, 9 partial scans of both sides of the figure were made. The number of partial scans was dictated by the need to increase the repeatability of scans due to the "difficult" surface of the object—glossy black stone with a small number of characteristic areas. Dealing with this challenge required maintaining a relatively uniform distance between the scanner and the object surface. Expecting the occurrence of glare and breaks in the scanned surface, the scanning of each figure side was repeated several times.

During the data processing stage, the obtained partial scans underwent processing and assessment. It lasted 2.5 h. As a result, one scan of unsatisfactory quality was rejected due to the occurrence of significant gaps and visible reflections on the saved texture. The remaining 8 scans were used to generate the high-quality base 3D model (Figure 7—on the left). The obtained model covered 100% of the "Snake" figure area—there were no areas that could not be digitized. The obtained mesh density allowed for a good reproduction of the object's shape and cavities (Figure 7—on the right).

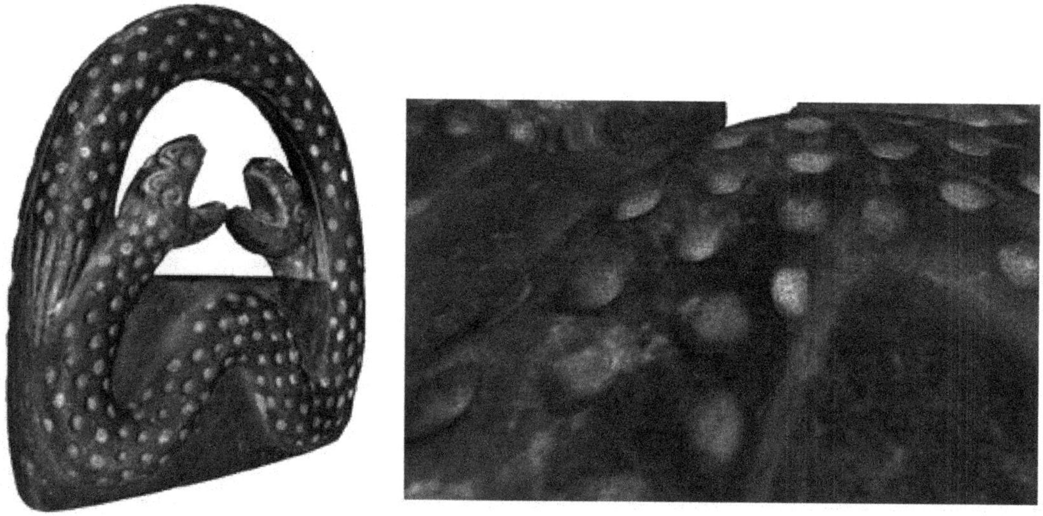

Figure 7. 3D base model of the "Snake" figure: overview (on the **left**) and mesh visible (on the **right**).

It took another 1 h to prepare the dissemination 3D model (Figure 8) for the purpose of publishing it on the 3D Digital Silk Road Portal [39]. The time spent is mainly the result of finding a balance between the model quality and providing fair performance for the viewers. The performance aspect was very important as the model was published within a larger collection on a website offering a dedicated 3D viewer. Other actions concerned polishing, converting and publishing the model to meet the requirements of the portal.

It was the simplest of the use cases. It can be seen from the shortest preparation, scanning and restoration times, which were no longer than 20 min. Due to the small dimensions and simple shape, the size of scans was the smallest, just as the processing time leading to the 3D base model. To conclude, in this case, the applied methodology turned out to be fully sufficient.

5.2. Case Study 2—"Icon of Saint Nicholas of Myra"

The icon was attached to the wall of the church main hall. During the accessibility assessment stage, the risk of moving the icon was evaluated as too high to be undertaken, thus it was decided to scan the icon from the front. It was possible without removing elements of the church's decor. The church administrators had no information about possible light sensitivity of the icon. Therefore, it was assumed that the permissible intensity of light cannot exceed the intensity of the church's own lighting. Scanning was carried out with the scanner's own lighting turned off. It turned out that such conditions were enough to obtain a satisfactory color of the texture. During the 12 min scan, four partial scans of the icon's surface and borders were obtained. Due to the variety of textures and colors, no repetition of scans was required.

Figure 8. 3D dissemination model of the "Snake" figure [40].

The processing stage confirmed the good quality of the obtained scans—all four scans were accepted and used in generation of the base model. (Figure 9—on the left). It lasted 1.5 h. The obtained base model is a representation of the icon's front part, which is the area of greatest importance. Despite the fact that only ~50% of the original object was scanned, it was sufficient for the object analysis and presentation. The back of the icon, assuming negligible or low historical value of its surface, could be digitally reconstructed, e.g., by means of CAD software. The obtained grid density and colors (Figure 9—on the right) allow for a good reproduction of the icon's surface appearance and texture. The dissemination model shows some discontinuities in the area of the upper frame (Figure 10), which are more strongly emphasized due to significant simplification of the 3D model mesh.

The less-than-perfect lighting conditions during the 3D scanning, chosen due to uncertainty about the icon vulnerability to strong lighting, triggered that the obtained final 3D model might slightly differ from the original in terms of colors. Nevertheless, the use of the 3DScaMOTO method allowed for creation of an acceptable base model (for archiving purposes) and a good dissemination model, while not exposing the object to unnecessary harm.

5.3. Case Study 3—"Buddha"

The surface of the statue with a uniform, non-glossy texture (limestone) made it possible to perform 3D scanning from the minimum allowable distance. It was important due to the limited space inside the display case.

Three dimensional scanning of the statue lasted 45 min in total. It took 35 min to scan the statue's front. 15 scans were made with different orientation and trajectory of the scanner in relation to the object. The aim was to obtain the fullest possible scan of the deep grooves (especially in the area depicting leaves of a Bodhi tree). In addition, five scans of the statue's rear surface were made, using the scanner placed outside the display case. Scanning that part took 10 min. Despite many attempts to scan with different scanner settings, it was impossible to perform a complete scan of the statue's front part—as expected (Figure 11).

Figure 9. 3D base model of the "Icon of Saint Nicholas of Myra": overview (on the **left**) and mesh visible (on the **right**).

Figure 10. 3D dissemination model of the "Icon of Saint Nicholas of Myra".

Figure 11. Incomplete 3D scan of the "Buddha" statue.

The side walls of the grooves in the area of a Bodhi tree leaves and the grooves around the monks have not been completely digitized. The problem here was not so much limited access to the object due to the display case, but the shape of the grooves: narrow, deep gaps smaller than the dimensions of the scanner. It was not possible to position the scanner so that it could properly see the problematic surfaces of the grooves with all its cameras. The statue's rear part, despite scanning through the glass, did not show similar problems (mainly due to a fairly uniform surface without major depressions and gaps). The scan matching process did not reveal any significant deformations due to potential interference while scanning through the glass.

During the data processing stage, the obtained partial scans underwent processing and assessment, which lasted 5.3 h. As a result, two scans of unsatisfactory quality were rejected. The remaining 18 scans were used to generate the high-quality base 3D model (Figure 12—on the left). It was estimated that the obtained model covered 85% of the original "Buddha" statue. Luckily, the obtained fragmentary scans of Bodhi tree leaves turned out to be sufficient to apply the algorithms for filling gaps in the meshes of triangles, obtaining the effect visually similar to the original (Figure 12—on the right). A similar situation took place in the areas around the monks, although this time the deviation from the original is more evident. Nevertheless, it was possible to obtain a complete 3D model of the statue, reproducing the original sufficiently in terms of shape and color—which was the main purpose of the scan (Figure 12—on the left).

It took another 1.2 h to prepare the dissemination 3D model (Figure 13) for the purpose of publishing it on the 3D Digital Silk Road Portal [39]—similarly as in the case of the "Snake" figure. Again, the time spent is mainly the result of finding a balance between the model quality and mesh complexity, thus providing fair performance for the viewers. Although this time the mesh was noticeably more complex.

It was the most complicated of the cases used. It can be seen from the long preparing, scanning and restoring times that lasted 70 min in total. Due to the dimensions, hard accessibility and troublesome shape of the surface influencing the scanning approach, the size of scans was large, just as the processing time leading to the 3D base model. To conclude, in this case, the methodology applied was verified successfully, despite the final model not fully complying with the original heritage object.

Figure 12. 3D base model of the "Buddha" statue: overview (on the **left**) and mesh visible (on the **right**).

Figure 13. 3D dissemination model of the "Buddha" statue [40].

6. Conclusions

Existing methodological approaches to 3D scanning of tangible heritage in most cases assume that the procedure of scanning is performed in places that are closed to tourists, at least for the time of scanning. However, closing an exhibition for tourists, or moving an artifact to be scanned, is not always possible.

The authors in this work proposed a methodological approach designed especially for small- and-medium sized objects to overcome difficulties expected in such cases. The procedure has been successfully implemented during the 3D scanning of objects of the Silk Road heritage as well as the wooden sacral architecture of the Maramures region. The authors provided three case studies, presenting different kinds of challenges, such as: an object being movable but of a glossy surface, an object being not movable but of good accessibility, at last an object being not movable and of bad accessibility. For each type of object, the procedural aspects were discussed, and the way from planning the scanning, through obtaining a 3D base model, to obtaining a 3D dissemination model was shown.

According to the authors, the research question posed in this work might be answered in the following way: "YES, it is possible to acquire, in an effective way, 3D scans using 3D scanners, specifically based on structured light, of historical objects while not removing these objects from their original area of exhibition.". This assumption is based on long-term experience of the authors in applying the proposed methodological approach, its implementation during the presented case studies and above all due to its results of use.

It was revealed that the proposed procedure successfully allowed to counteract organizational problems during 3D scanning of heritage objects being under continuous display, and that the achieved results of scanning were nevertheless of good quality. In the case of the "Snake" figure, while applying the method, the goals of digitization were fully achieved. In the case of the "Buddha" statue, the digitization outcome's deficiencies were not caused by the shortcomings of the method itself, but by limitations of the equipment used. Filling gaps in the dissemination model, and leaving gaps in the base model, might be the temporary solution. The case of the "Icon of Saint Nicholas of Myra" is similar, the difference lying in the need of CAD-like reconstruction of the icon's back, due to the inability to move it from the church wall.

What is more, based on the authors' experience it could be concluded that:

1. Digitization using a hand-held structured-light 3D scanner is quite a simple process, that could be performed on heritage sites without the need of providing extensive facilitation for the scanning persons.
2. Digitization using structured-light technology is safe for heritage objects, as there is no need for sticking markers on the object's surface, as well as the objects are not warmed up thanks to the technology being low energy.
3. The 3D scanning technology described in this work, allows to capture photorealistic textures. As a result, the photographic documentation made by a digital camera does not have to be transformed to 3D model textures. The time and effort during postprocessing are saved.
4. High precision of scans allows the creation of high-quality 3D models for the heritage object archaization purposes. Such models might later be used, e.g., for reconstruction purposes, as dimensions of the original object might be read from them. Such models might also be adjusted according to dissemination purposes.

Further works concerning 3D digitization of heritage objects being under continuous display will be conducted in compliance with the method presented in this work. This should further verify the method and be the source of valuable insights for the method modifications.

Author Contributions: Conceptualization, J.K. and K.Ż.; methodology, J.K., M.M., K.Ż. and J.M.; software, J.K.; validation, J.K., M.M., J.M., M.J. and C.N.; formal analysis, K.Ż. and J.K.; investigation, J.K., K.Ż., M.M. and J.M.; resources, J.K., J.M., M.J. and C.N.; data curation, J.K., M.M. and M.J.; writing—original draft preparation, K.Ż., J.K. and J.M.; writing—review and editing, K.Ż., M.J. and M.M.; visualization, K.Ż. and J.K.; supervision, J.K., M.M. and K.Ż.; project administration, M.M., K.Ż. and J.K.; funding acquisition, M.M. and J.M. All authors have read and agreed to the published version of the manuscript.

Funding: The APC was funded by the Lublin University of Technology Scientific Fund FD-22/IT-3/999.

Institutional Review Board Statement: Not applicable.

Informed Consent Statement: Not applicable.

Data Availability Statement: Not applicable.

Acknowledgments: The authors thank Jannat Ismailova (State Museum of History of Uzbekistan, Tashkent, Uzbekistan).

Conflicts of Interest: The authors declare no conflict of interest.

References

1. New European Bauhaus. Available online: https://europa.eu/new-european-bauhaus/about/about-initiative_en (accessed on 9 September 2022).
2. EREA-EU-Funded Cultural Heritage Projects. Available online: https://rea.ec.europa.eu/news/eu-funded-cultural-heritage-projects-2022-05-03_en (accessed on 9 September 2022).
3. EREA-€264 Million Available to Finance Research on Cultural Heritage Preservation, Migrant Integration, Democracy and the Creative Industries. Available online: https://rea.ec.europa.eu/news/eu264-million-available-finance-research-cultural-heritage-preservation-migrant-integration-2022-01-20_en (accessed on 9 September 2022).
4. UNDP-Social and Environmental Standards, Standard 4. Available online: https://info.undp.org/sites/bpps/SES_Toolkit/SitePages/Standard%204.aspx (accessed on 9 September 2022).
5. Montusiewicz, J.; Miłosz, M.; Kęsik, J.; Żyła, K. Structured-light 3D scanning of exhibited historical clothing—A first-ever methodical trial and its results. *Herit. Sci.* **2021**, *9*, 74. [CrossRef]
6. Miłosz, M.; Montusiewicz, J.; Kęsik, J.; Żyła, K.; Miłosz, E.; Kayumov, R.; Anvarov, N. Virtual scientific expedition for 3D scanning of museum artifacts in the COVID-19 period—The methodology and case study. *Digit. Appl. Archaeol. Cult. Herit.* **2022**, *26*, e00230. [CrossRef]
7. Miłosz, M.; Skulimowski, S.; Kęsik, J.; Montusiewicz, J. Virtual and interactive museum of archaeological artefacts from Afrasiyab—An ancient city on the silk road. *Digit. Appl. Archaeol. Cult. Herit.* **2020**, *18*, e00155. [CrossRef]
8. Skublewska-Paszkowska, M.; Powroźnik, P.; Smołka, J.; Miłosz, M.; Łukasik, E.; Mukhamedova, D.; Miłosz, E. Methodology of 3D scanning of intangible cultural heritage—The example of Lazgi dance. *Appl. Sci.* **2021**, *11*, 11568. [CrossRef]
9. Żyła, K.; Kęsik, J.; Santos, F.; House, G. Scanning of historical clothes using 3D scanners: Comparison of goals, tools, and methods. *Appl. Sci.* **2021**, *11*, 5588. [CrossRef]
10. Comes, R.; Neamțu, C.; Buna, Z.L.; Bodi, S.; Popescu, D.; Tompa, V.; Ghinea, R.; Mateescu-Suciu, L. Enhancing accessibility to cultural heritage through digital content and virtual reality: A case study of the Sarmizegetusa Regia UNESCO site. *J. Anc. Hist. Archaeol.* **2020**, *7*, 124–139. [CrossRef]
11. Popovici, D.-M.; Iordache, D.; Comes, R.; Neamțu, C.G.D.; Băutu, E. Interactive exploration of virtual heritage by means of natural gestures. *Appl. Sci.* **2022**, *12*, 4452. [CrossRef]
12. Gomes, L.; Regina, O.; Bellon, P.; Silva, L. 3D reconstruction methods for digital preservation of cultural heritage: A survey. *Pattern Recognit. Lett.* **2014**, *50*, 3–14. [CrossRef]
13. Bruno, F.; Bruno, S.; De Sensi, G.; Luchi, M.-L.; Mancuso, S.; Muzzupappa, M. From 3D reconstruction to virtual reality: A complete methodology for digital archaeological exhibition. *J. Cult. Herit.* **2010**, *11*, 42–49. [CrossRef]
14. Sampaio, A.Z.; Gomes, A.M.; Sánchez-Lite, A.; Zulueta, P.; González-Gaya, C. Analysis of BIM methodology applied to practical cases in the preservation of heritage buildings. *Sustainability* **2021**, *13*, 3129. [CrossRef]
15. Pepe, M.; Costantino, D.; Alfio, V.S.; Restuccia, A.G.; Papalino, N.M. Scan to BIM for the digital management and representation in 3D GIS environment of cultural heritage site. *J. Cult. Herit.* **2021**, *50*, 115–125. [CrossRef]
16. Miłosz, M.; Kęsik, J.; Montusiewicz, J. 3D scanning and visualization of large monuments of Timurid architecture in Central Asia—A methodical approach. *J. Comput. Cult. Herit.* **2021**, *14*, 1–31. [CrossRef]
17. Daneshmand, M.; Helmi, A.; Avots, E.; Noroozi, F.; Alisinanoglu, F.; Arslan, H.S.; Gorbova, J.; Haamer, R.E.; Ozcinar, C.; Anbarjafari, G. 3D scanning: A comprehensive survey. *arXiv* **2018**, arXiv:1801.08863v1. [CrossRef]
18. Miłosz, M.; Montusiewicz, J.; Kęsik, J. *3D Information Technology in the Protection and Popularization of the Cultural Heritage of the Silk Road*; Wydawnictwo Politechniki Lubelskiej: Lublin, Poland, 2022; ISBN 978-83-7947-521-6. Available online: http://bc.pollub.pl/dlibra/publication/14114/edition/13771 (accessed on 9 November 2022).
19. Montusiewicz, J.; Czyż, Z.; Kayumov, R. Selected methods of making three—Dimensional virtual models of museum ceramic objects. *Appl. Comput. Sci.* **2015**, *11*, 51–65.
20. Kowalski, M.; Naruniec, J.; Daniluk, M. LiveScan3D: A fast and inexpensive 3D data acquisition system for multiple Kinect v2 sensors. In Proceedings of the 2015 International Conference on 3D Vision, Lyon, France, 19–22 October 2015. [CrossRef]
21. Tucci, G.; Bonora, V.; Conti, A.; Fiorini, L. High-quality 3D models and their use in a cultural heritage conservation project. *Int. Arch. Photogramm. Remote Sens. Spat. Inf. Sci.* **2017**, *XLII-2/W5*, 687–693. [CrossRef]
22. Ciekanowska, A.; Kiszczak-Gliński, A.; Dziedzic, K. Comparative analysis of Unity and Unreal Engine efficiency in creating virtual exhibitions of 3D scanned models. *J. Comput. Sci. Inst.* **2021**, *20*, 247–253. [CrossRef]
23. Salwierz, A.; Szymczyk, T. Methods of creating realistic spaces—3D scanning and 3D modelling. *J. Comput. Sci. Inst.* **2020**, *14*, 101–108. [CrossRef]
24. Akça, D.; Grün, A.; Breuckmann, B.; Lahanier, C. High definition 3D-scanning of arts objects and paintings. In Proceedings of the Optical 3-D Measurement VIII, Zurich, Switzerland, 9–12 July 2007; Volume II, pp. 50–58.
25. Wei, O.C.; Chin, C.S.; Majid, Z.; Setan, H. 3D documentation and preservation of historical monument using terrestrial laser scanning. *Geoinf. Sci. J.* **2010**, *10*, 73–90.
26. Neamtu, C.; Comes, R.; Popescu, D. Methodology to create digital and virtual 3D artefacts in archaeology. *J. Anc. Hist. Archeol.* **2016**, *3*, 65–74. [CrossRef]

27. Bianco, G.; Gallo, A.; Bruno, F.; Muzzupappa, M. A comparison between active and passive techniques for underwater 3D applications. *Int. Arch. Photogramm. Remote Sens. Spat. Inf. Sci.* **2011**, *XXXVIII-5/W16*, 357–363. [CrossRef]
28. Bianco, S.; Ciocca, G.; Marelli, D. Evaluating the performance of structure from motion pipelines. *J. Imaging* **2018**, *4*, 98. [CrossRef]
29. Barszcz, M.; Montusiewicz, J.; Paśnikowska-Łukaszuk, M.; Sałamacha, A. Comparative analysis of digital models of objects of cultural heritage obtained by the "3D SLS" and "SfM" methods. *Appl. Sci.* **2021**, *11*, 5321. [CrossRef]
30. 3D Scanning of Exhibits from the County Museum in Nysa. Available online: https://global3d.pl/en/module/psblog/module-psblog-blog?id=77 (accessed on 19 November 2022).
31. Girelli, V.A.; Tini, M.A.; Dellapasqua, M.; Bitelli, G. High resolution 3D acquisition and modelling in cultural heritage knowledge and restoration projects: The survey of the Fountain of Neptune in Bologna. *Int. Arch. Photogramm. Remote Sens. Spat. Inf. Sci.* **2019**, *XLII-2/W11*, 573–578. [CrossRef]
32. Kęsik, J.; Miłosz, M.; Montusiewicz, J.; Samarov, K. Documenting the geometry of large architectural monuments using 3D scanning—The case of the dome of the Golden Mosque of the Tillya-Kori Madrasah in Samarkand. *Digit. Appl. Archaeol. Cult. Herit.* **2021**, *22*, e00199. [CrossRef]
33. Kvítková, Z.; Petru, Z. Challenges of tourism management in cultural UNESCO sites in the V4 countries from sustainability perspective. In Proceedings of the 23rd International Conference on Environmental Economics, Policy and International Environmental Relations, Prague, Czech Republic, 25–26 November 2021; pp. 102–107, ISBN 978-80-7490-230-7.
34. Kvítková, Z.; Petru, Z.; Houška, P.; Macáková, L. Impact of the inscription of the cultural landscape on the UNESCO list on tourism destination. In Proceedings of the 25th International Colloquium on Regional Sciences, Brno, Czech Republic, 22–24 June 2022; pp. 375–381. [CrossRef]
35. Kvítková, Z.; Petru, Z.; Šauer, P. Sustainability in smaller UNESCO sites from the entrepreneurs' perspective. In Proceedings of the Aktuální Problémy Cestovního Ruchu, Jihlava, Czech Republic, 2–3 March 2022; pp. 1–12.
36. Šauer, P.; Dvořák, A.; Mildeová, S.; Mokrišová, J. Vyjádření užitku přírodního statku metodou podmíněného hodnocení: Případ snížení rizika záplav. *Politická Ekon.* **1998**, *46*, 412–426. [CrossRef]
37. Sauer, P.; Fiala, P.; Dvorak, A.; Kolinsky, O. Coalition projects in wastewater treatment: The case of a drinking water reservoir in the Czech Republic. *J. Environ. Prot. Ecol.* **2015**, *16*, 1492–1501.
38. Popa, A. Biserica de lemn din Păușa—Sălaj. *Acta Musei Napoc.* **1971**, *XVIII*, 593–602.
39. Silk Road 3D Portal—Home. Available online: https://silkroad3d.com (accessed on 9 November 2022).
40. Silk Road 3D Portal—Selected artefacts of the National Museum of the History of Uzbekistan. Available online: https://silkroad3d.com/wp-content/web3DView/tashkent1/?desclang=EN (accessed on 9 November 2022).

applied sciences

MDPI

Article

Investigating a Workflow for Obtaining Physical Models from Digital Twins Obtained through Photogrammetry and TLS: New Ways for a Sustainable Dissemination of Heritage

Pablo Alejandro Cruz Franco [1,*], Adela Rueda Márquez de la Plata [2] and María Pérez Sendín [2]

[1] Department of Construction, Polytechnic School, University of Extremadura, 10003 Cáceres, Spain
[2] Department of Graphic Expression, Polytechnic School, University of Extremadura, 10003 Cáceres, Spain
* Correspondence: pablocruzfranco@unex.es; Tel.: +34-927257195

Abstract: The objective of this article is to generate and validate a workflow that allows us to print physical twins (models) from digital twins obtained from unique buildings of our architectural heritage. These physical twins will guarantee the preservation and diffusion of the cultural asset, and will promote new pedagogical ways to spread the heritage. In addition, these physical cufflinks are an interesting tool to guarantee universal accessibility through low-cost and rapid distribution typological models. We can overcome physical barriers and reach many more users, regardless of their economic, physical, or location conditions. To obtain this workflow, digital twins of architectural elements of high cultural value obtained by photogrammetry or terrestrial laser scanner will be used. The digital twin will be optimized through different platforms, and an optimized exchange file will be generated for its dissemination and printing. A digital platform will be proposed that guarantees the free distribution of these digital twins to any user. Finally, the physical twins will be obtained. For the development of this work, we have chosen to use 3D resin printers with SLA technology (selective laser exposure to light by laser) due to their performance and high quality in the models obtained.

Keywords: physical twins; 3D print; digital heritage; digital survey; 3D data base; photogrammetry; VAT polymerization; SLA

Citation: Cruz Franco, P.A.; Rueda Márquez de la Plata, A.; Pérez Sendín, M. Investigating a Workflow for Obtaining Physical Models from Digital Twins Obtained through Photogrammetry and TLS: New Ways for a Sustainable Dissemination of Heritage. *Appl. Sci.* **2023**, *13*, 1057. https://doi.org/10.3390/app13021057

Academic Editors: Marek Milosz and Jacek Kęsik

Received: 11 December 2022
Revised: 27 December 2022
Accepted: 5 January 2023
Published: 12 January 2023

1. Introduction

This research develops a methodology to facilitate the dissemination, conservation, and universal accessibility of our architectural heritage through physical prototyping from digital twins.

In this research, we have two clearly differentiated stages: the first, which we could define as "from the physical to the virtual", that is, the capture or digitization of reality through new technologies, and the second stage: from the digital to the physical, that is, the physical prototyping from the data obtained or, in other words, the printing using 3D technologies of the captured data (see Figure 1).

- This article establishes the following three objectives: Assess and standardize a workflow that allows us to arrive from a high-resolution «digital twin» (output formats: obj, 3ds, stl, etc.) obtained through photogrammetry [1,2] or TLS [3] to a physical model printed using current technologies [4,5].
- Obtain replicable models of our heritage that serve to bring our monuments closer to the greatest possible number of people through "maker" technologies.
- Promote the "capture" of these "digital models/digital twins" and their free distribution as good practices to guarantee the conservation and dissemination of heritage.
- Try to promote a philosophy of free distribution that brings our heritage closer to as many people as possible through these physical prototypes and their exchange files.

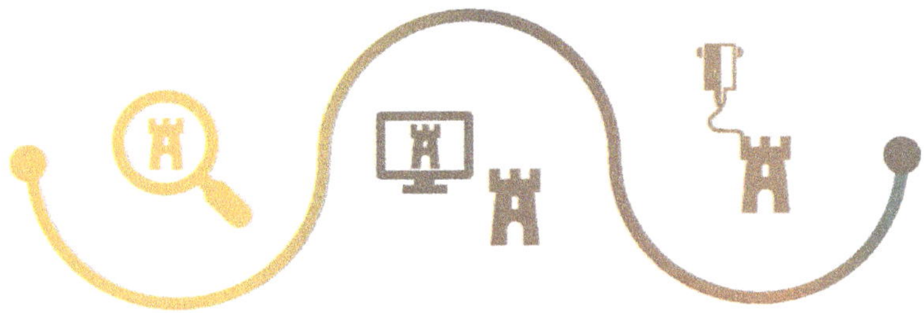

Figure 1. Labor process: "from the physical to the digital and anchor from the digital to the physical": (**a**) Capture of the physical monument using new technology (**b**) Elaboration to generate digital gemelli of our heritage (**c**) Obtain physical gemelli from new technology.

As a three-dimensional working model to test this methodology, we used the digital twin made for the «restoration project for the wall of Cáceres Phase II. Almohad towers of the southwestern wall in Cáceres». This project has been carried out by the architects Isabel Bestué Cardiel (team director), Carmen Cañones Gallardo, Rosario Carmona Campos, Pablo Alejandro Cruz Franco, Adela Rueda Márquez de la Plata, and Javier Chavez Quesada. (See Figure 2).

This restoration project, of which the authors of this document have been a part, has been developed on the Almohad walls of the world heritage city of Cáceres [6,7]. The work area is divided into two sections. The first section, known as the Adarve del Padre Rosalío section, is made up of a section of wall parallel to the Adarve del Padre Rosalío street, an albarrana tower, and a defensive cube. This section has a floor area of 170.79 m^2.

The second section of the wall has an intrasdos with an extension of 16.62 linear metres. The section is on urban land and public space, and has a total floor area of 155 m^2. Through the extrados of this section, a double section of wall can be identified, interrupted by a flanking tower to the southeast. Beneath this stretch of wall, a possible antemural rises from Calle Cornudilla. At the corner of Calle Adarve del Padre Rosalío there is an albarrano wall and an octagonal albarrana tower.

The capture of this digital twin, as we have said, has been carried out using photogrammetry. Work has been carried out by combining point clouds obtained by SfM (with images acquired from UAV, Unnamed Aerial Vehicle) [8,9]. To verify the reliability of the model, the models have been contrasted with the point clouds obtained by means of a digital instrument, a Terrestrial Laser Scanner, TLS [1,10–12], and the result is a three-dimensional model. This is a high quality canvas model [6] of the city, and we can affirm that it has been obtained through a low-cost and precise "intelligent work" flow, having been developed essentially through photogrammetry and verifying the reliability of the model through TLS. All of this digital twin has been processed in cabinet work [13].

When we work with three-dimensional models made up of point clouds and triangulated models, a key point is to validate the geometric quality of our models. As we have said, the model obtained through SfM guarantees a very complete model, thanks to its ability to reach almost all points, especially if we compare it with a model obtained through TLS. On the other hand, this model may contain geometric inaccuracies that must be verified, so the model obtained through TLS helps us to verify the quality of the model through SfM. On the one hand, this model obtained through TLS is less complete due to the physical limitation of access to roofs, which causes phantom zones, but, on the other hand, it is a much more reliable model in terms of precision, since we know the millimeter precision and it helps us to control and validate the results. Ground photogrammetry was more accurate, but has the same limitation as three-dimensional models obtained by TLS. Aerial photogrammetry is less accurate, but reduces the number of ghost zones to almost 0.

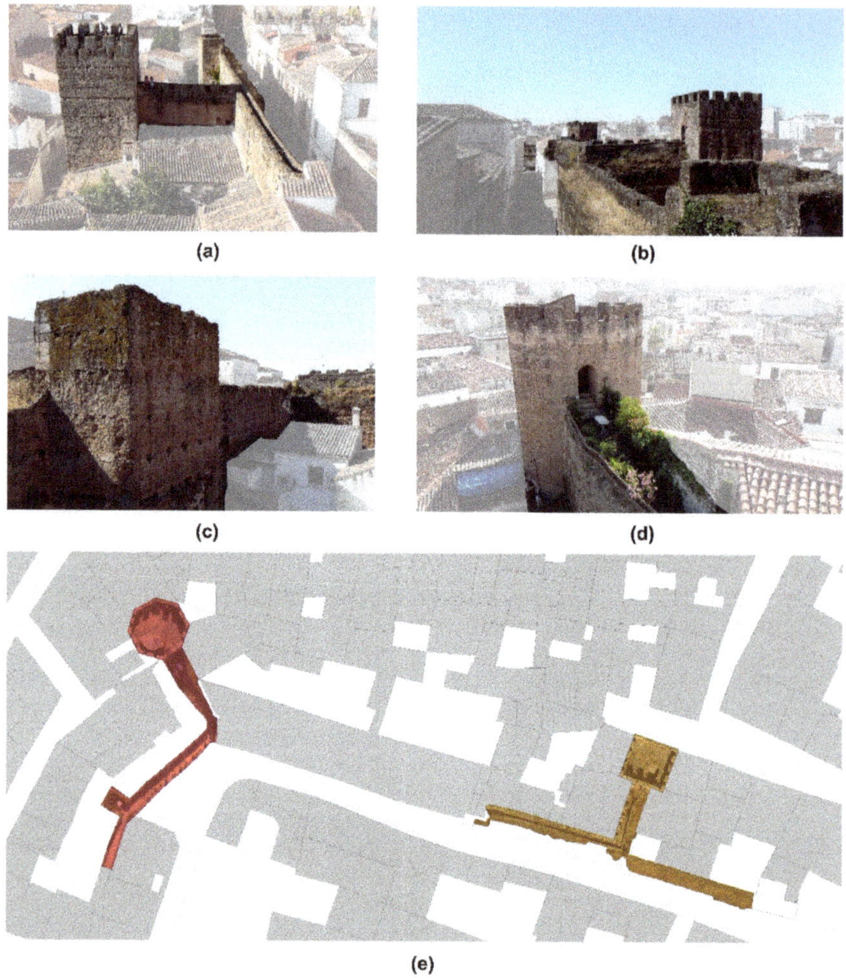

Figure 2. (**a**) Aerial view of the Aver tower, (**b**) view of the fortified complex from the intrados of the canvas. In the foreground tower of the Aver, (**c**) cube of the Aver. Trasdós of the canvas, (**d**) aerial view of the Ochavada tower, (**e**) plan of the study area. Right canvas, colored in red, cube and Ochavada tower. Left canvas, colored in orange, cube and tower of the Aver.

In the geometric validation procedure, we obtain two point clouds. To scale and orient them, we used the RECAP platform to facilitate and guarantee the transition to BIM (Building Information Modeling), in our case, on the REVIT platform, since both belonged to the Autodesk set of tools, where interoperability was guaranteed.

The point cloud obtained using Agisoft Metashape is in a *.psx format. The *.psx format is not compatible with RECAP, so we need a previous step in which we modify the format of our file to a format suitable for RECAP. In the case of the cloud obtained through TLS, there is no problem since we use the cyclone software to process the data obtained through the BLK360 laser, which allows us to export and resample our point clouds in formats compatible with RECAP.

In this way, the cloud of points was obtained in Metashape. To do this, we exported from .psx to .laz. The .laz format, as we have said, is compatible with the RECAP program and allows us to load that cloud, process it, classify the points, etc. Once these accessions have been made, the next step is to link the cloud in REVIT.

The next step will be carried out within the REVIT platform. With the point cloud in RECAP format, we will proceed to link it in REVIT. We will call this cloud of points 1 A. We can have as many as we want, for example, 1 A, 1 B, 1 C, etc. Using homologous points, we will proceed to position these clouds in the model space, making them overlap.

According to what was seen in the article "Protocol Development for Point Clouds, Triangulated Meshes and Parametric Model Acquisition and Integration in an HBIM Workflow for Change Control and Management in a UNESCO's World Heritage Site" [1], we will proceed to make a chromatic differentiation of each of the point clouds to easily visualize geometric differences.

We will proceed to an analysis through sections, elevations, and plans at different levels. This analysis is carried out quickly in REVIT, allowing us to measure and obtain fast and reliable data about the quality of our models, being able to discard entire models or parts of them if necessary. With this system, the level of precision is analyzed correctly.

2. Related Works

2.1. Obtaining the Digital Twin

Obtaining these digital twins is a recurring theme in the scientific literature that deals with the digital survey of heritage. It is important to highlight the DADA Lab of the University of Pavia, directed by Professor Sandro Parrinello, and the LS3D Lab of the University of Florence, directed by Professor Stefano Bertocci, due to their importance, and for being pioneers in this area. Both laboratories are an international benchmark in the field of architectural survey, documentation, and protection of heritage [2,14–21].

To obtain the digital twin [13,22,23] a system is necessary that allows us to carry out different flight missions at different times in order to capture the entire study area [8]. This system has to be versatile, not affect assets, and be low cost. It is cabinet work; this work represents a great advance and innovation in itself, respecting the current state of cultural heritage.

After the study is carried out, it has been decided to use a work methodology, consisting of the creation of a more exhaustive "own" polygonal mesh that allows us to georeference the study area with greater accuracy, facilitating the realization of different flight missions at different times; this work represents a great advance and innovation in itself while respecting the current state of cultural heritage (see Figure 3).

To generate the polygonal mesh, it is necessary to insert some nails (metal spikes) into the pavement that allow us to place the exact targets/targets; once they are located at the points of interest, we proceed to georeferencing with respect to the city simulating topographical nails.

Once we have a flight program and defined the structure of the missions and their objectives, the data will be collected using UAV [8,20,24–26].

For an exact geometric definition, a TLS (Terrestrial Laser Scanning) will be used [27] that will allow us to accurately measure the monument and check the accuracy of the shots made with UAV [2] (see Figure 4).

Thanks to TLS technology we quickly and accurately obtain a three-dimensional model with possible phantom zones, such as the roofs of a building, but by combining these with the models generated from UAV technology, we can obtain complete models of great precision that will allow us to later process the models to generate the physical prototypes [7,27].

This network is susceptible to expansion over time, making it possible to have a series of points that can be used to establish a new network as a result of deterioration, loss, movement of its original position due to inclement weather, vandalism, works, and incidents on the pavement, to link future work models or to carry out different survey campaigns.

In order to interconnect future three-dimensional models or the different flight missions with the use of three nails (points), it is already possible to reference the model, but to work more safely and accurately in this work, more points have been used, thus facilitating complementary work. Once the three-dimensional model has been obtained, it can be expanded by linking the targets of the previous work with new ones closer to the area on which it is intended to work, and so on, until the entire scope of the project is covered, that is, an extensive project is sectioned into smaller areas for more detailed and precise work. The different models can be linked, obtaining the necessary information by using complementary techniques after the creation of the mesh, such as UAV (Unnamed aerial vehicle) photogrammetry, SLR (Single lens reflex) or the TLS (Terrestrial Laser Scanning) model [28].

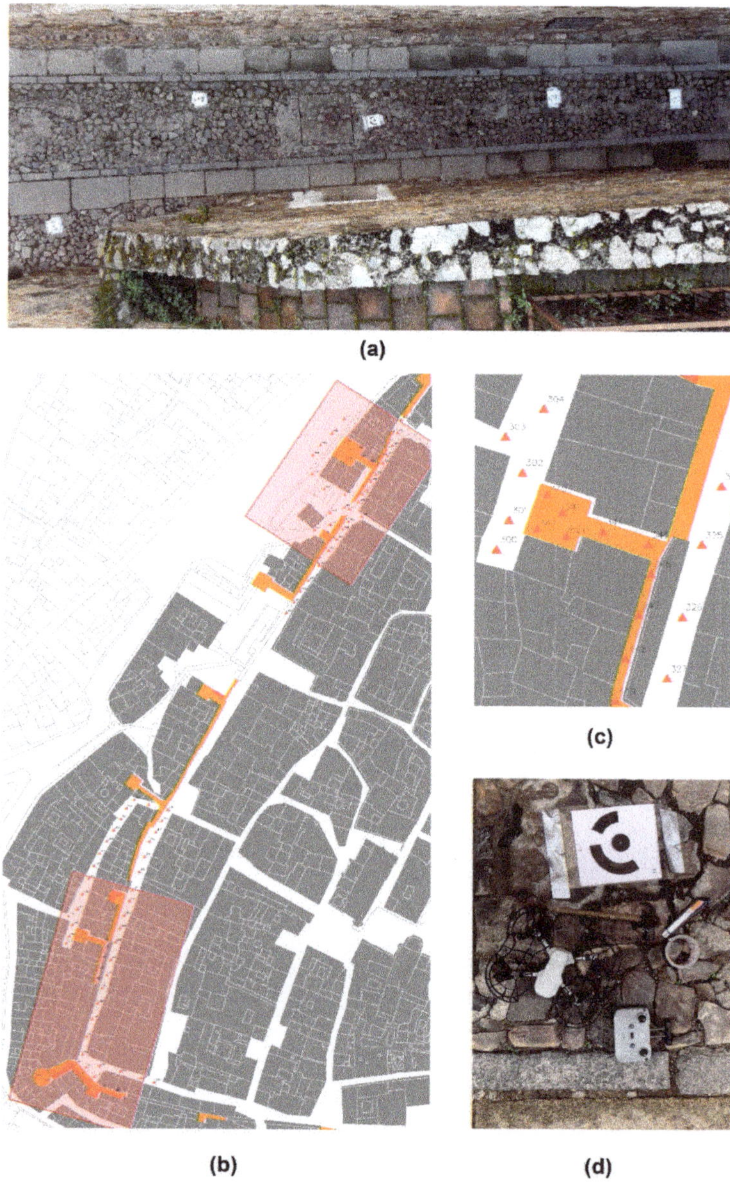

Figure 3. (**a**) Polygonal mesh including the targets, (**b**) work areas and positioning of the different points that make up the mesh. (**c**) Detail of the work area corresponding to the Aver tower (**d**) Work elements.

Figure 4. In the image: (**Left**): the author flying the drone. (**Right**): the author positioning the laser scanner.

2.2. Obtaining the Physical Prototype

Today, the way of understanding reality in our society is constantly evolving given the ease of access to the digital tools that we currently have. In addition, the use of new technologies has been increased, given the situation that COVID-19 has generated, producing a great change when it comes to relating to each other [29–32] and with the heritage that surrounds us [33–36]. The need arises to bring heritage closer to any part of the world in the simplest way possible; therefore, the need for digital twins has become relevant and, going a little further, combining digital twins with the most avant-garde technologies related to 3D printing, we can obtain physical prototypes that allow us to discover the cultural and architectural heritage in a different and cheaper way [11,25,37,38].

We can say that there are three large families of 3D printing technologies. In the first place, we would say that we have the family of addition (additive manufacturing), that is, the one that encompasses all the technologies that replicate reality by adding material, where nothing existed. Based on layers, we can generate our physical twin; in a certain way, we could compare it to the Greek technique of casting with lost wax [26,39]. Secondly, we have subtraction technologies (subtractive manufacturing); these technologies, from blocks of raw material, by means of milling cutters or other elements, remove layers of material and give rise to our physical twin as if it were a classic stone sculpture (normally we will talk about CNC machines here). Lastly, we would talk about cutting technologies (manufacturing by cutting): these technologies, based on sheets of material and cutting templates (DWG, DXF, etc.), will generate pieces that we will assemble; normally, these are laser cutting machines, although we could generate similar parts with CNC printers. In Figure 5, we can see different examples developed by the research team in collaboration with the University of the Americas of Puebla, México.

Figure 5. Different examples of physical models of architecture obtained using different 3D printing and cutting systems. TAD3 Lab. Works in collaboration between the University of the Americas of Puebla and the University of Extremadura. In the figure, we can see models made of wood on both sides that have been executed with a laser printer and CNC cutting machine, and, in the center of the table, we can see white filament models that have been manufactured with a filament extrusion printing printer.

In the case of this research, it has been decided to use an addition technology, more specifically, VAT polymerization, specifically laser SLA (selective exposure to light by laser). A brief summary of the different additive manufacturing technologies has been prepared below:

First of all, if we see Figure 6a, we have the material extrusion technology FDM (fused deposition modelling). This technology stands out for its optimal surface finishes, a wide range of colors and materials, and a low cost. On the contrary, we have low resolution and, in many cases, fragility of the pieces. Secondly, in Figure 6b, we have outlined the operation of the technology known as material jetting (MJ). This procedure allows the creation of objects by combining different materials and colors. It is based on photopolymeric liquid resin that is heated to 30–60 °C, thus obtaining the right viscosity for printing. The print head, which is very similar to that of an inkjet printer, distributes the photopolymer resin on the platform. The curing of this resin is performed with UV light. The projector of this UV light is integrated into the print head and will form different superimposed layers. Like the SLA process that we will see later, this technology is based on the principle of photopolymerization. Among its characteristics is an optimal surface finish, a full range of colors, and a multitude of materials with different physical and mechanical properties, but it has a high cost. Thirdly, in Figure 6c, within the additive printing technologies, we highlight the selective laser sintering (SLS) technology, which is a rapid prototyping technique in which a layer of powder is deposited (a few tenths millimeter) in a vat that has been heated to a temperature slightly below the melting point of the powder.

Next, a high-power laser causes the deposited particles to fuse and solidify. This process is called sintering. With SLS technology, we obtain functional parts with good mechanical properties, and we can make complex geometries. Among its negative characteristics would be that it has long delivery times and a higher cost than FDM technology. In Figure 6d, we see a scheme of the sand or metallic powder printing system, known as 3D binder jetting. This technique consists of the deposition of an adhesive binding agent on thin layers of powder material. Powder materials are metal (e.g., stainless steel) or ceramic (e.g., glass or plaster) based. This technology is low cost, and we can generate functional metal parts. Lastly, in Figure 6e, we have the latest addition printing technology studied: VAT polymerization. This family brings together the technologies that achieve 3D printing through the polymerization of a vat (VAT) filled with photosensitive resin.

Depending on the light source used, we will have different classes: laser SLA (selective exposure to light by laser), DLP-SLA (selective exposure to light by projector), and MSLA (selective exposure to light masked by LCD). All of them will obtain a final item obtained by accumulating layers.

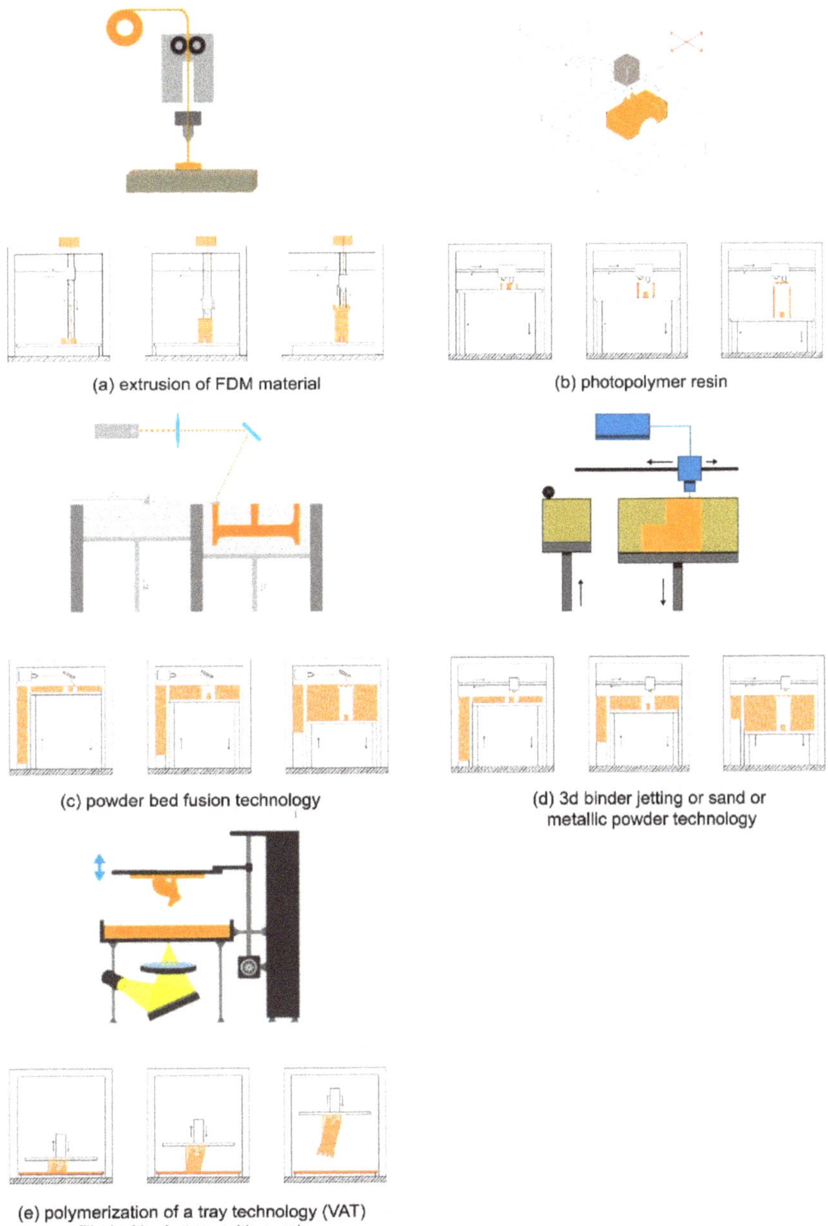

Figure 6. Three-dimensional printing technologies: (**a**) printing through the extrusion of FDM material; (**b**) photopolymer resin (**c**) powder bed fusion technology; (**d**) 3D binder jetting or sand or metallic powder technology; (**e**) polymerization of a tray technology (VAT) filled with photosensitive resin.

2.3. Choice of Printing Technology

As we have indicated, the project is developed in the City of Cáceres, a World Heritage Site, and its physical development will be developed within the FAD LAB: SOL (Smart Open Lab), located in the Polytechnic School of Cáceres at the University of Extremadura.

SOL is an open production space that has supported this project with tools, machines, and knowledge of the community that composes it. SOL's main philosophy is to share knowledge globally and, thus, contribute to development and research.

The printing possibilities within SOL are multiple. Once the resolution of the obtained models and the machines at our disposal were analyzed, it was decided to use the equipment: Form 2. This equipment allows us to obtain physical models using photopolymer resin and SLA laser technology (selective exposure to light by laser). Specifically, the form 2 printer allows us to obtain exceptional resolution and precision in details, specifically between 50 microns and 100 microns.

3. Results: A Standardized Protocol in Five Steps

After carrying out the process, we have managed to generate a standardized protocol with five general steps as detailed in Figure 7.

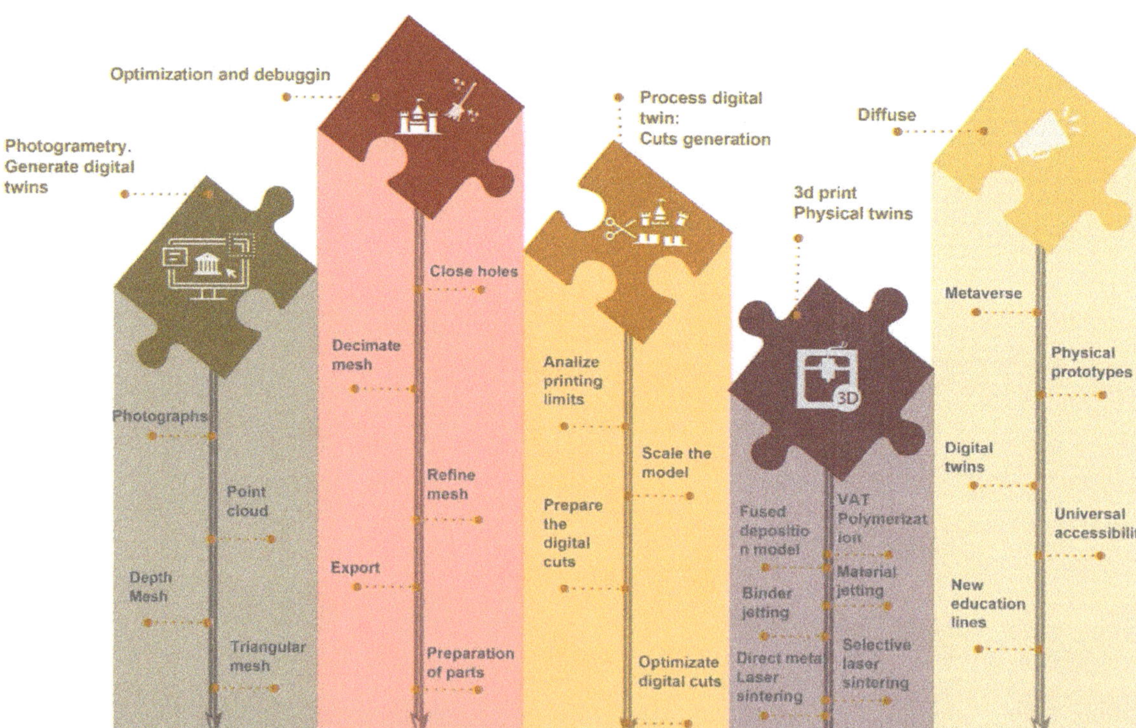

Figure 7. Five steps of the standardized protocol.

3.1. Generation of Digital Twins

As we have indicated, the development of this digital twin has been carried out by combining point clouds obtained by SfM (with images acquired from UAV) and clouds obtained by TLS digital instruments; the result is a 3D digital model [40,41]. It is a high-quality model of the city canvas, and we can say that it has been obtained through a low-cost and precise "smart work" flow [13]. The steps to obtain this physical model are summarized in Figure 8 [1,7,25,42–44]:

- Orientation of photographs: this orientation allows the scattered point cloud to be obtained.
- Generation of dense point cloud and depth maps.
- Dense point cloud optimization by point filtering.
- Generation of the triangulated mesh.
- Generation of the texture from the images taken by the drone.
- This processing gives rise to digital twins made up of huge meshes, which need to be reviewed and refined to obtain three-dimensional models that can be used in metaverses [13,45–48].

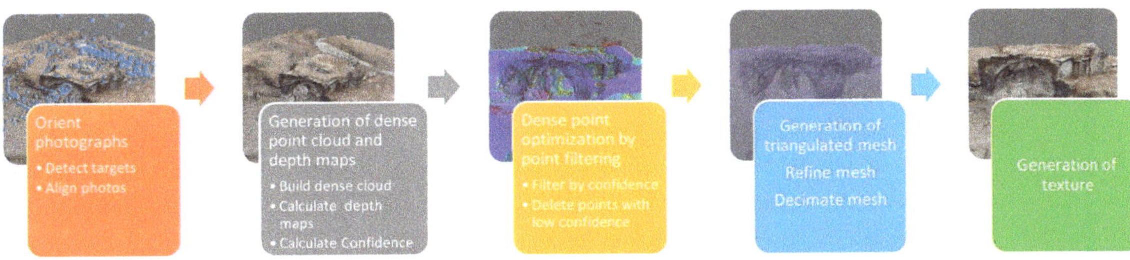

Figure 8. Photogrammetric processing of the data obtained by UAV.

In the case of the model obtained with TLS, we obtain a second virtual twin, which will be built from the documentation using TLS. This is cabinet work (see Figure 9).

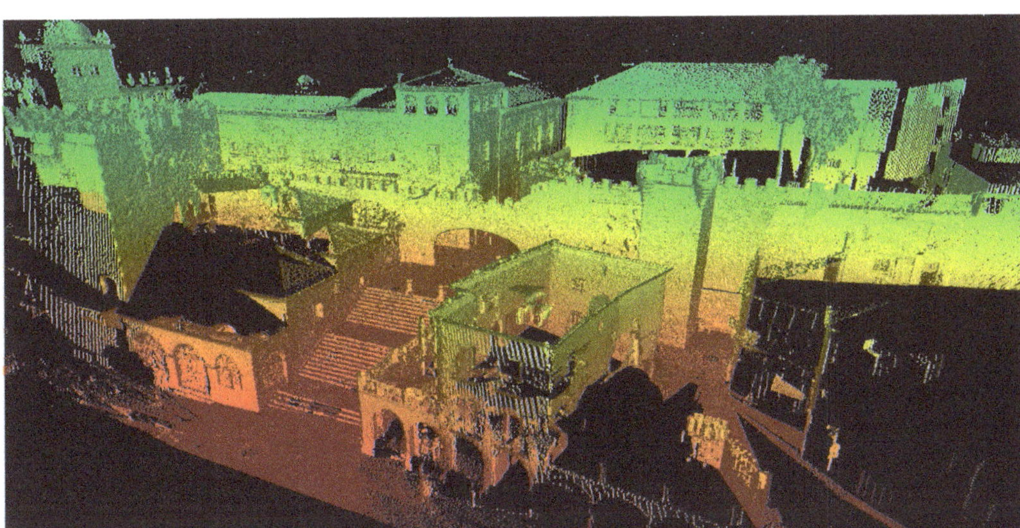

Figure 9. Results of the point cloud obtained with TLS.

After the study is carried out, it has been decided to use the software of the laser scanner Cyclone 360 register, which is software for photogrammetric processing of digital images and generation of 3D spatial data. Its main functions include: automatic alignment of scans, manual alignment of the scans according to the point clouds, allowing the link with other point clouds through the use of targets, and texturing options through the images generated in the scans (see Figure 10):

- Step 1: Importing Scans
- Step 2: Aligning Scans

- Step 3: Model optimization
- Step 4: Detect targets or common and fixed points
- Step 5: 2D planimetry from the point cloud
- Step 6: Export the point cloud for further processing.

Figure 10. TLS processing of the data obtained.

3.2. *Optimization and Debugging of the 3D Models Obtained to Obtain Exportable Digital Twins*

As we have indicated in the previous section, the three-dimensional model obtained through photogrammetry and TLS [49,50] is a complex and heavy model that requires a lot of computational load [13]. It becomes a priority to optimize it to obtain exportable files in different exchange formats. To debug the model, we use Rhinoceros and Blender interchangeably. With these programs we can repair and reduce the mesh [51].

The first step is to repair the mesh, as it is imperfect, and it is necessary to close holes derived from bad data collection in areas where the point cloud is not dense enough to form a closed mesh from it. The next thing is to reduce the mesh, that is, reduce the number of triangles, until arriving at a simplified mesh that considerably reduces the size of the source file. Then we smooth the mesh. Mesh smoothing consists of approximating straight edges to radii with a given curvature while maintaining mesh density.

At this point, we have mainly used three types of software that have allowed us to study the different models and compare them to search for errors and to suppress ghost zones. Mainly the Rhino software, Blender and the instant meshes application have been used. We have made a retopology of the models and closed holes, as we have commented. We have gone from a triangular mesh to a quadrangular mesh, and we have repositioned the high quality textures in the corrected model using Blender; they are not necessary for the model printed in TLS, but for digital twins uploaded to repositories for AR and VR if necessary.

3.3. *Processing of the Digital Twins: Generation of Cuts*

This third processing step of this «digital twin» has been carried out again using the Rhinoceros 3D software. We have focused on the environment of the wall, and we have modeled the semi-detached houses in a simplified way. Once this first phase of modeling has been completed, it has been divided into fragments according to the physical limit imposed by the printer model that we have used (form 2), which is 145 × 145 × 175 mm. In total, 24 cuts have been planned, indicated in Figure 11.

In the scheme, we also see a blue area that represents the environment a little further away from buildings. This environment was made using the laser cutter. It was simplified from the parcel of the Historic District.

Below we leave, the Table 1 summarizes the different pieces generated, related to the printing time and the material necessary for its execution. In total, 24 cuts have been planned, indicated in the scheme that we see below. These 24 cuts require 1907.83 mL of white resin (not counting the printing tests), and two resin tank units have also been used. A total of 203 h have been invested in printing alone. These data do not take into account the cleaning and curing time (carried out with the formwash and formcure machines, also from 3dlab, as we will see later) or the time spent on polishing, sanding, etc. (carried out in a traditional carpentry workshop).

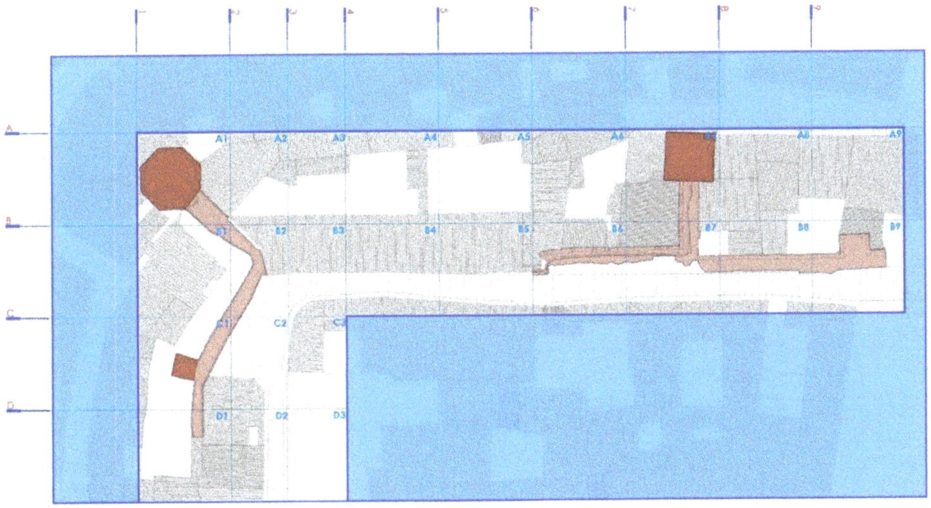

Figure 11. Plan of planning and organization of cuts and work according to the selected printer model.

Table 1. Summary table of physical prototype.

Pieces	Swap File (STL)	Preform	Resin (mL)	Duration (h)
A1	A1_STL	A1_FORM	114.95 mL	10 h
A1	A1_STL	A1_FORM	96.63 mL	10 h 15 min
A2	A2_STL	A2_FORM	77.96 mL	7 h 45 min
A3	A3_STL	A3_FORM	62.31 mL	7 h 15 min
A4	A4_STL	A4_FORM	74.02 mL	8 h 15 min
A5	A5_STL	A5_FORM	68.26 mL	7 h
A6	A6_STL	A6_FORM	92.40 mL	9 h 15 min
A7	A7_STL	A7_FORM	78.64 mL	9 h 45 min
A8	A8_STL	A8_FORM	76.91 mL	7 h 45 min
A9	A9_STL	A9_FORM	70.23 mL	6 h 45 min
B1	B1_STL	B1_FORM	72.46 mL	7 h 30 min
B2	B2_STL	B2_FORM	49.63 mL	6 h 45 min
B3	B3_STL	B3_FORM	50.63 mL	7 h 15 min
B4	B4_STL	B4_FORM	79.17 mL	9 h 30 min
B5	B5_STL	B5_FORM	87.84 mL	10 h 15 min
B6	B6_STL	B6_FORM	280.31 mL	16 h 30 min
B7	B7_STL	B7_FORM	76.31 mL	9 h 30 min
B8	B8_STL	B8_FORM	76.56 mL	9 h 15 min
B9	B9_STL	B9_FORM	79.89 mL	9 h 30 min
C1	C1_STL	C1_FORM	84.07 mL	8 h 45 min
C2	C2_STL	C2_FORM	49.13 mL	6 h 30 min
C3	C3_STL	C3_FORM	51.25 mL	7 h 15 min
D1	D1_STL	D1_FORM	90.15 mL	9 h
D2	D2_STL	D2_FORM	55.26 mL	7 h
D3	D3_STL	D3_FORM	27.81 mL	4 h 30 min

3.4. 3D Print

As we have indicated, this work is carried out within the FAD Lab SOL (Smart Open Lab) and a FORM 2 unit has been selected to make the impressions. A versatile workflow is generated that allows the transfer of the information obtained, thanks to photogrammetry (digital twins) to twins [4] high resolution physics (see Figure 12).

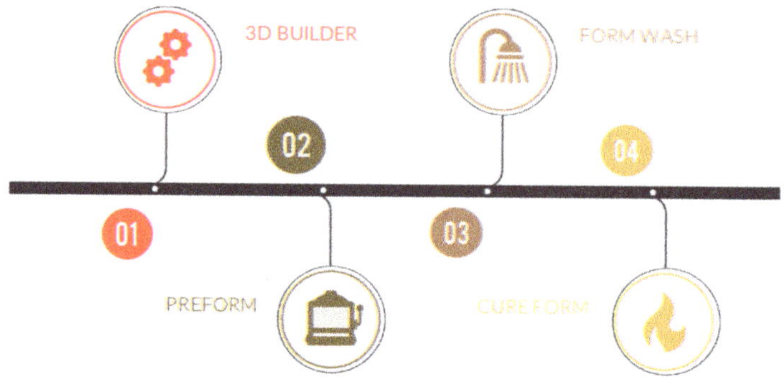

Figure 12. Diagram of the process for 3D printing.

3.4.1. Phase 1. Export of the Photogrammetric Model

Nowadays, there are many ways of exchanging digital files for the different software that we use. In our case, we exported the model with the OBJ extension, since it allows us to keep the textures of our digital twin.

3.4.2. Phase 2. Processed in 3D Builder

To process the digital model, we used the 3D Builder software, free software from the Microsoft Corporation company, that allows users to create, model, and print models in a generalized and free way, being a great help to facilitate the transmission of knowledge in any part (see Figure 13).

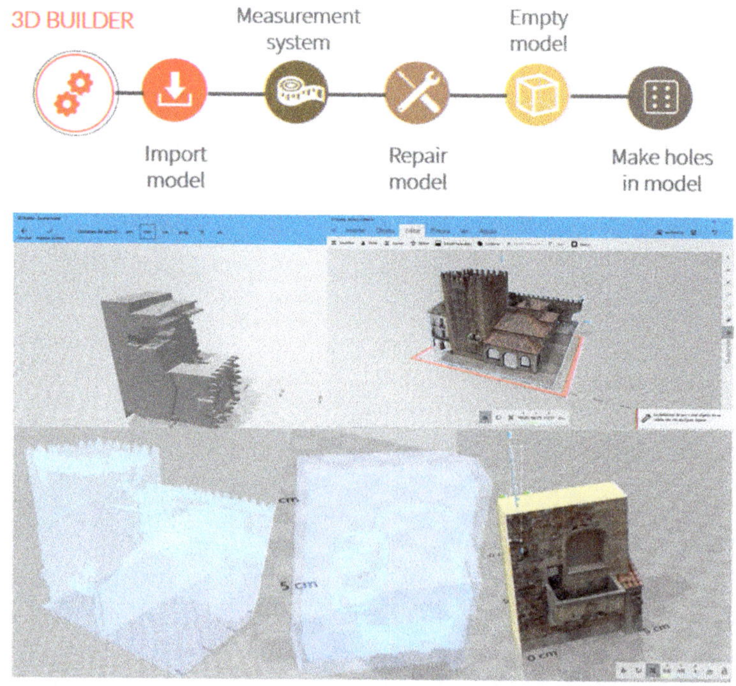

Figure 13. Workflow for processing digital twins in 3D Builder.

3.4.3. Phase 3. Preform Processing

Once the digital twin has been processed, it is necessary to continue processing it, this time with Preform, the free software of the Form 2 printer, from the company Form Labs, which uses advanced parameters to generate the supports and optimize the printing of each piece, resulting in a digital model ready for printing (see Figure 14a).

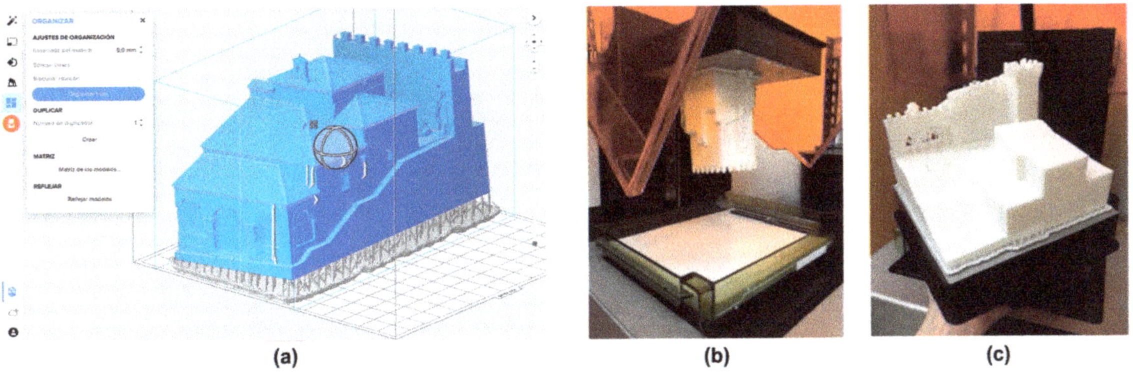

Figure 14. (**a**) digital twin before printing (**b**,**c**) physical prototype on the Form 2 printer.

3.4.4. Phase 4. Printing

To do this, the following equipment has been used within the Smart Open Lab: 3D printer, specifically, the Form 2 model (a stereolithography (SLA) 3D printer), Form Wash, and Cure Form.

FormLabs, Form 2: It works with a resin printer due to the high quality it offers (50–100 microns) (see Figure 14b,c).

Washing in Form Wash: Once the pieces are obtained, it is necessary to wash them to remove the excess resin without the surface curing. This the machine uses isopropyl alcohol, and the sink time will depend on the type of resin that is being used.

Cured in Cure Form: Once the parts have been washed, the supports that have been generated prior to printing are removed. Once the pieces are without supports, they are cured; for this, it is necessary to use the Cure Form. The curing time and temperature will depend on the type of resin that is being used for the prototypes.

3.4.5. Phase 5. Assembly

Once the different pieces of the model have been obtained, both in resin (obtained with the Form 2 printer) and in DM board (obtained with the laser cutter), in specific cases there may be joints that we have had to review. These joints can be treated as wood to improve the assembly, always keeping in mind that these areas are together; they are not the faces of the model obtained by TLS [52] or photogrammetry, in which we have that aforementioned precision of microns and that could not be touched by these methods. In these specific cases, it should be noted that the behavior of the resin used has been very similar to that of wood, in that it can be sanded, cut, drilled, etc., expanding the possibilities of interaction with this material and the possibilities of executing objects. See Figure 15.

3.4.6. Phase 6. Diffusion

This point becomes a labor capital section. We have taken the following path: from the physical to the digital and from the digital to the physical. That is to say, that first moment, in which the material object that is the element of architectural heritage, at the moment in which we have identified that it is a work of art and decided to digitize it for posterity; and that second moment, in which the digitized work becomes universal, and we fight to reach all corners of the world in that search for universal accessibility.

Figure 15. Process of assembly and assembly of the physical prototype.

In this way, we are going to have two diffusion lines. First of all, the model itself, obtained by our research group using SLA laser technology (selective exposure to light by laser). In Figure 16, we can see the final results and the work shots. Secondly, as we have said, the work would be philosophically serious; the vocation of this work is the free distribution of this knowledge in order to generate an international knowledge matrix. For this, the use of different platforms has been proposed [13].

Figure 16. Finalist prototype and taking of previous photographs.

As we have said within the working group, different platforms have been studied that are used to house digital twins. We highlight the result of the article presented in the journal Applied Sciences with the name: Protocols for the Graphic and Constructive Diffusion of Digital Twins of the Architectural Heritage That Guarantee Universal Accessibility through AR and VR [13,53]. Among the results obtained, we highlight the following two platforms: Thingiverse (www.thingiverse.com, accessed on 2 December 2022) and Sketchfab (www.sketchfab.com, accessed on 15 December 2022). The first, thingiverse, is a free online website dedicated to the creation and publication of digital STL files, designed and published by the users that make up the community. When a community member decides to upload a file, they will be able to choose the type of license they want to add to it. Second, the Sketchfab Platform is an outstanding platform, due to the great variety of content that it incorporates, as well as its 3D model viewer. Sketchfab allows downloading of files from the web if the users (owners of the models) allow it. We want to highlight the VR functionalities provided by this second platform.

4. Discussion

We are at a time of change, in which the tools that are made available to the architect to protect heritage are immeasurable. These tools allow us to generate extensive 3D databases that allow heritage protection from knowledge.

In this communication, we present a workflow for the generation of three-dimensional models and their dissemination. These workflows generate new possibilities that extend the concept of universal accessibility. We want to highlight the following possibilities:

First of all, in this expanded concept of universal accessibility, the situation arises in which anyone, regardless of their economic situation or where they live, can enjoy these cultural elements.

In second place, in a teaching level, it is a step forward in the tools that we provide to teachers. All these resources are already interactive, as we have said, and AR, VR [13], or physical are revolutions in the classroom, which that allow a teacher from anywhere in the world to print a wall like the one at hand for their students, so that they can study and document it.

In our experience as professors at a university of technical education (architecture), the student's training in construction subjects improves when they have construction details in their hands. Incorporating printing technologies as activities in teaching have very profitable results by incorporating transversal skills into teaching that substantially improve the abilities of students (See Figure 17).

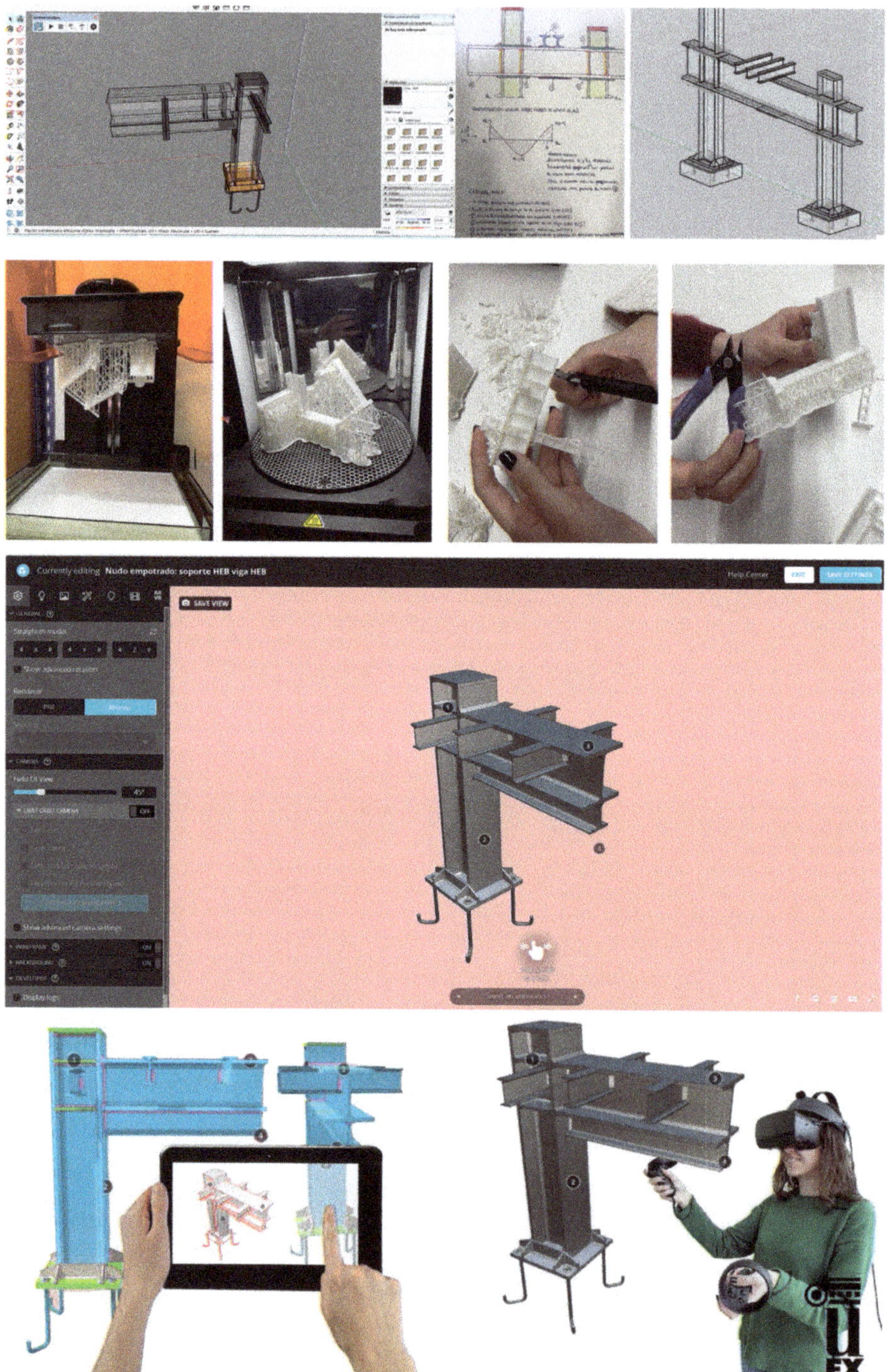

Figure 17. Work process within the classroom experiences at the University of Extremadura with construction students. Prototyping process, generation of physical models, uploading these models to the open platform sketch fab, and immersive experiences with students in VR.

Thirdly, considering open digital repositories such as Sketchfab or Thingverse is a very interesting altruistic proposal for universities around the world. Indeed, 3D printing cannot replace reality, but it is an aid for the teacher that, as we have been saying, breaks the barriers of space and money. It is a way of offering new interactive resources to our teachers and students.

Fourthly, all of these 3D model databases can also be expanded in that aforementioned universal accessibility to blind people, guaranteeing an open repository at a global level [54].

Three-dimensional printing technology is already an undeniable reality in our society. This "advanced science" has made its way over the last few years and, little by little, the "maker" culture has transformed into a fresh and energetic movement that is revolutionizing many areas of our lives through its ideas. This culture is changing the way we understand teaching, architecture, or engineering through new proposals for dealing with design, new methods of understanding prototyping, and new ways of motivating new generations [55].

These original proposals to motivate current students are a challenge in teaching because they propose alternatives to traditional teaching, specifically, in this case, teaching under construction. These alternatives are neither better nor worse; they are always complementary to the traditional ones, but absolutely necessary as transversal training for future technicians.

The incorporation of these activities into teaching is very beneficial, as it incorporates cross-cutting skills into teaching that substantially improve students' abilities: these new abilities will help our technicians in future academic and vital projects.

To each time its art, and to each art its freedom
(«Der Zeit ihre Kunst, der Kunst ihre Freiheit»)
(Gustav Klint, friso de entrada al edificio de la Secesión vienesa, 1897–1898)

5. Conclusions

The proposed methodology is divided into two phases. The first phase is data collection and the generation of a physical twin. Note that this first phase has been extensively developed in other bibliographies that are cited during the text, and in other publications of the research group to which we refer. Regarding the first phase, we state that:

- First of all, as a society, we have the obligation to digitize the world in which we live in order to preserve it as a digital twin for future generations: a database of the present.
- Secondly, we have the opportunity to recreate from these digital twins lost architectural environments and future digital environments: foundations of the past and the future.
- There are many digitization methods, each with its advantages and disadvantages [4,17,26,56–58]. In the present study, a low-cost methodology based on the use of UAVs has been chosen [8,59], which has been outlined and incorporated into the workflow. This scheme is supported by other works and the professional experience of the authors [1,25].
- For some time now, new representation technologies have allowed us to create resources and content that reflect how our heritage is and could be. These digital twins, both the digital copies of the present and the digital proposals of the past and future, give rise to enormous databases that have traditionally been used for the development of restoration projects, such as the one studied at the Hermitage of Vaquero, but once the work was completed, it was complex and difficult to implement as a knowledge tool.
- The proposed methodology allows us to increase our capacity to divulge, protect, and conserve our heritage. Digital borders are crossed, and we obtain prototypes that can be distributed in open source to promote knowledge regardless of place, physical, or economic conditions.
- On the other hand, these proposals represent a revolution at the educational level; our students can have in their hands replicas of the monuments with millimetric quality.

At present, education is in a permanent change. It is necessary to incorporate all these novelties to improve the skills of our teachers and our students [60].

- On the other hand, these physical prototypes are a tool that opens the doors to that aforementioned universal accessibility through typological models of our heritage.
- As has been seen in the discussion of results, this way of working can be extrapolated in different subjects, such as teaching, construction, architecture, pathologies, history, etc.

In short, all of this article illustrates a new system for the generation of physical prototypes that will allow us to conserve art in new ways, and allow us to improve our teaching and outreach tools.

"Beauty perishes in life but is immortal in art"—
Leonardo Da Vinci.

Author Contributions: The editorial responsibility of the paragraphs is recognized to: P.A.C.F., A.R.M.d.l.P. and M.P.S.; Conceptualization, P.A.C.F. and A.R.M.d.l.P.; methodology, P.A.C.F. and M.P.S.; validation P.A.C.F. and M.P.S.; formal analysis A.R.M.d.l.P. and P.A.C.F.; investigation P.A.C.F., A.R.M.d.l.P. and M.P.S.; resources, P.A.C.F. and A.R.M.d.l.P.; writing—original draft preparation, P.A.C.F. and M.P.S.; writing—review and editing, P.A.C.F., A.R.M.d.l.P. and M.P.S.; visualization, P.A.C.F., A.R.M.d.l.P. and M.P.S.; supervision, P.A.C.F. and M.P.S.; project administration, P.A.C.F. and A.R.M.d.l.P. All authors have read and agreed to the published version of the manuscript.

Funding: This publication has been made possible thanks to funding granted by the Consejería de Economía, Ciencia y Agenda Digital de la Junta de Extremadura and by the European Regional Development Fund of the European Union through the reference grant GR21159 (COMPHAS researcher group). In addition, thanks to funding granted by Consejería de Economía, Ciencia y Agenda Digital de la Junta de Extremadura and by the European Regional Development Fund of the European Union through the Researcher project "Aplicación de tecnologías VR y levantamientos 6D para la implementación de una accesibilidad universal en el patrimonio arqueológico de edificación pública romana" through the reference grant IB20096. For the development of this research, we have had the support of the Comphas research group of the University of Extremadura, which has provided us with the material means and tools, we have been able to put into practice the proposed working method.

Institutional Review Board Statement: Not applicable.

Informed Consent Statement: Not applicable.

Data Availability Statement: Not applicable.

Acknowledgments: This publication has been developed within the TAD3 Investigation group of the University of Extremadura, which is currently developing virtual tours of the heritage of the region of Extremadura, we have managed to disseminate this heritage asset. Furthermore, it would not have been possible without the support of the FabLab SOL (Smart Open Lab) based at the Polytechnic School of Cáceres or its members. A very special thanks to Antonio Gordillo Guerrero and Mario Figueira Torres for the training they have given us in the use of the machinery and for having provided all the necessary machinery for the development of the prototypes. Either for the development of this research, we have had the support of the Comphas research group of the University of Extremadura, which has provided us with the material means and tools, we have been able to put into practice the proposed working. This publication has been partially developed thanks to the SEXPE Innovation and Talent Program 2021 of the Junta of Extremadura: Protocolos para la implementación de modelos de información en ciudades patrimonio de la humanidad para la búsqueda de una accesibilidad universal And finally, very special thanks to Paula Pérez Sendín for her help and unconditional collaboration in this work.

Conflicts of Interest: The authors declare no conflict of interest.

References

1. Rueda Marquez de la Plata, A.; Cruz Franco, P.A.; Cruz Franco, J.; Gibello Bravo, V. Protocol Development for Point Clouds, Triangulated Meshes and Parametric Model Acquisition and Integration in an HBIM Workflow for Change Control and Management in a UNESCO's World Heritage Site. *Sensors* **2021**, *21*, 1083. [CrossRef] [PubMed]
2. Parrinello, S.; Picchio, F. Integration and Comparison of Close-Range Sfm Methodologies for the Analysis and the Development of the Historical City Center of Bethlehem. *ISPRS—Int. Arch. Photogramm. Remote Sens. Spat. Inf. Sci.* **2019**, *42*, 589–595. [CrossRef]
3. Tan, J.; Leng, J.; Zeng, X.; Feng, D.; Yu, P. Digital Twin for Xiegong's Architectural Archaeological Research: A Case Study of Xuanluo Hall, Sichuan, China. *Buildings* **2022**, *12*, 1053. [CrossRef]
4. Pérez Sendín, M. *Prototipado Físico a Partir de Gemelos Digitales Aplicado a la Torre de Bujaco*; Universidad de Extremadura: Cáceres, Spain, 2022.
5. Niccolucci, F.; Felicetti, A.; Hermon, S. Populating the Data Space for Cultural Heritage with Heritage Digital Twins. *Data* **2022**, *7*, 105. [CrossRef]
6. Cruz Franco, J.; Cruz Franco, P.A.; Rueda Márquez de la Plata, A.; Ramos Rubio, J.A.; Rueda Muñoz de San Pedro, J.M. Evolución histórica de la muralla de Cáceres y nuevos descubrimientos. El postigo de San Miguel, un lienzo perdido de la cerca almohade. *Rev. De Estud. Extrem.* **2016**, *Tomo LXXII, Número III*, 1869–1910.
7. Cruz Franco, P.A.; Rueda Márquez de la Plata, A.; Cruz Franco, J.; Ramos Rubio, J.A. A lost Fragment and Gate of the Almohad Wall of the World Heritage City of Cáceres, Spain. *IOP Conf. Ser. Mater. Sci. Eng.* **2017**, *245*, 052051. [CrossRef]
8. Gómez Bernal, E.; Cruz Franco, P.A.; Rueda Márquez de la Plata, A. Drones in architecture research: Methodological application of the use of drones for the accessible intervention in a roman house in the Alcazaba of Mérida (Spain). In Proceedings of the D-SITE Drines-Systems of Information on culTural hEritage, Pavia, Italy, 24–26 June 2020.
9. Templin, T.; Popielarczyk, D. The Use of Low-Cost Unmanned Aerial Vehicles in the Process of Building Models for Cultural Tourism, 3D Web and Augmented/Mixed Reality Applications. *Sensors* **2020**, *20*, 5457. [CrossRef]
10. Kushwaha, S.K.P.; Dayal, K.R.; Sachchidanand; Raghavendra, S.; Pande, H.; Tiwari, P.S.; Agrawal, S.; Srivastava, S.K. 3D Digital Documentation of a Cultural Heritage Site Using Terrestrial Laser Scanner—A Case Study. In *Applications of Geomatics in Civil Engineering*; Springer: Singapore, 2020; pp. 49–58.
11. Cruz Franco, P.A.; Rueda Márquez de la Plata, A.; Cruz Franco, J. From the Point Cloud to BIM Methodology for the Ideal Reconstruction of a Lost Bastion of the Cáceres Wall. *Appl. Sci.* **2020**, *10*, 6609. [CrossRef]
12. Żyła, K.; Kęsik, J.; Santos, F.; House, G. Scanning of Historical Clothes Using 3D Scanners: Comparison of Goals, Tools, and Methods. *Appl. Sci.* **2021**, *11*, 5588. [CrossRef]
13. Cruz Franco, P.A.; Rueda Márquez de la Plata, A.; Gómez Bernal, E. Protocols for the Graphic and Constructive Diffusion of Digital Twins of the Architectural Heritage That Guarantee Universal Accessibility through AR and VR. *Appl. Sci.* **2022**, *12*, 8785. [CrossRef]
14. Parrinello, S.; Francesca, P.; Dell'Amico, A.; De Marco, R. Prometheus. protocols for information models libraries tested on heritage of upper kama sites. msca rise 2018. In Proceedings of the II Simposio UID di Internazionalizzazione Della Ricerca. Patrimoni Culturali, Architettura, Paesaggio e Design tra Ricerca e Sperimentazione Didattica, Firenze, Italy, 22 October 2019.
15. Parrinello, S.; Gomez-Blanco, A.; Picchio, F. *El Palacio Del Generalife: Del Levantamiento Digital Al Proyecto de Gestión: Cuaderno de Trabajo Para la Documentación Arquitectónica*; Pavia University Press, Ed.; Pavia University Press: Pavia, Italy, 2017; Volume 1, p. 224.
16. Parrinello, S.; Morandotti, M.; Valenti, G.; Piveta, M.; Basso, A.; Inzerillo, A.; Lo Turco, M.; Picchio, F.; Santagati, C. *Digital & Documentation: Databases and models for the enhancement of Heritage*; Parrinello, S., Ed.; Edizioni dell'Università degli Studi di Pavia: Pavia, Italy, 2019; Volume 1.
17. Parrinello, S.; Picchio, F.; De Marco, R.; Dell'Amico, A. Documenting the Cultural Heritage Routes. The Creation of Informative Models of Historical Russian Churches on Upper Kama Region. *ISPRS—Int. Arch. Photogramm. Remote Sens. Spat. Inf. Sci.* **2019**, *XLII-2/W15*, 887–894. [CrossRef]
18. Parrinello, S.; Porzilli, S. Rilievo Laser Scanner 3D per l'analisi morfologica e il monitoraggio strutturale di alcuni ambienti inseriti nel progetto di ampliamento del complesso museale degli Uffizi a Firenze. In Proceedings of the Reuso 2016: Contributi per la Documentazione, Conservazione e Recupero Del Patrimonio Architettonico e Per la Tutela Paesaggistica, Pavia, Italy, 6–7 October 2016.
19. Parrinello, S.; Cioli, F. Un progetto di recupero per il complesso monumentale di Usolye nella regione della Kama Superiore. *Restauro Archeol.* **2018**, *26*, 92–111.
20. Bertocci, S.; Arrighetti, A.; Bigongiari, M. Digital Survey for the Archaeological Analysis and the Enhancement of Gropina Archaeological Site. *Heritage* **2019**, *2*, 848–857. [CrossRef]
21. Parrinello, S.; Dell'Amico, A. Experience of Documentation for the Accessibility of Widespread Cultural Heritage. *Heritage* **2019**, *2*, 1032–1044. [CrossRef]
22. Cruz Franco, P.A.; Rueda Marquez de la Plata, A.; Cortés Pérez, J.P. *Veinte Siglos de Patrimonio de Extremadura en Ocho Puentes. Documentación Digital de Las Obras Públicas*; Junta de Extremadura; Consejería de Economía e Infraestructuras y Universidad de Extremadura: Extremadura, Spain, 2018; Volume 1, p. 208.
23. Farella, E.M.; Morelli, L.; Rigon, S.; Grilli, E.; Remondino, F. Analysing Key Steps of the Photogrammetric Pipeline for Museum Artefacts 3D Digitisation. *Sustainability* **2022**, *14*, 5740. [CrossRef]

24. Almukhtar, A.; Saeed, Z.O.; Abanda, H.; Tah, J.H.M. Reality Capture of Buildings Using 3D Laser Scanners. *CivilEng* **2021**, *2*, 214–235. [CrossRef]
25. Ramos Sánchez, J.A.; Cruz Franco, P.A.; Rueda Márquez de la Plata, A. Achieving Universal Accessibility through Remote Virtualization and Digitization of Complex Archaeological Features: A Graphic and Constructive Study of the Columbarios of Merida. *Remote Sens.* **2022**, *14*, 3319. [CrossRef]
26. Arapakopoulos, A.; Liaskos, O.; Mitsigkola, S.; Papatzanakis, G.; Peppa, S.; Remoundos, G.; Ginnis, A.; Papadopoulos, C.; Mazis, D.; Tsilikidis, O.; et al. 3D Reconstruction & Modeling of the Traditional Greek Trechadiri: "Aghia Varvara". *Heritage* **2022**, *5*, 1295–1309. [CrossRef]
27. Saura-Gómez, P.; Spairani-Berrio, Y.; Huesca-Tortosa, J.A.; Spairani-Berrio, S.; Rizo-Maestre, C. Advances in the Restoration of Buildings with LIDAR Technology and 3D Reconstruction: Forged and Vaults of the Refectory of Santo Domingo de Orihuela (16th Century). *Appl. Sci.* **2021**, *11*, 8541. [CrossRef]
28. Andriasyan, M.; Moyano, J.; Nieto-Julián, J.E.; Antón, D. From Point Cloud Data to Building Information Modelling: An Automatic Parametric Workflow for Heritage. *Remote Sens.* **2020**, *12*, 1094. [CrossRef]
29. Lee, H.; Hwang, Y. Technology-Enhanced Education through VR-Making and Metaverse-Linking to Foster Teacher Readiness and Sustainable Learning. *Sustainability* **2022**, *14*, 4786. [CrossRef]
30. Lee, U.-K. Tourism Using Virtual Reality: Media Richness and Information System Successes. *Sustainability* **2022**, *14*, 3975. [CrossRef]
31. Okanovic, V.; Ivkovic-Kihic, I.; Boskovic, D.; Mijatovic, B.; Prazina, I.; Skaljo, E.; Rizvic, S. Interaction in eXtended Reality Applications for Cultural Heritage. *Appl. Sci.* **2022**, *12*, 1241. [CrossRef]
32. Shehade, M.; Stylianou-Lambert, T. Virtual Reality in Museums: Exploring the Experiences of Museum Professionals. *Appl. Sci.* **2020**, *10*, 4031. [CrossRef]
33. Zhao, S. Creating Futuristic Heritage Experiences: An Exploratory Co-Design Study through Design Fiction. *Sustainability* **2022**, *14*, 12027. [CrossRef]
34. Zhang, L.; Li, M.; Zhang, L.; Liu, X.; Tang, Z.; Wang, Y. MasterSu: The Sustainable Development of Su Embroidery Based on Digital Technology. *Sustainability* **2022**, *14*, 7094. [CrossRef]
35. Alves-Cardoso, F.; Campanacho, V. To Replicate, or Not to Replicate? The Creation, Use, and Dissemination of 3D Models of Human Remains: A Case Study from Portugal. *Heritage* **2022**, *5*, 1637–1658. [CrossRef]
36. Papagrigorakis, M.J.; Maravelakis, E.; Kyparissi-Apostolika, N.; Stravopodi, E.; Konstantaras, A.; Apostolikas, O.; Toulas, P.; Potagas, C.; Papapolychroniou, T.; Mastoris, M.; et al. An Integrated Study of the Mesolithic Skeleton in Theopetra Cave, Greece: From the Skeleton Analysis to 3D Face Reconstruction. *Heritage* **2022**, *5*, 881–895. [CrossRef]
37. Rueda Márquez de la Plata, A.; Cruz Franco, P.A.; Ramos Sánchez, J.A. Architectural Survey, Diagnostic, and Constructive Analysis Strategies for Monumental Preservation of Cultural Heritage and Sustainable Management of Tourism. *Buildings* **2022**, *12*, 1156. [CrossRef]
38. Salem, T.; Dragomir, M. Options for and Challenges of Employing Digital Twins in Construction Management. *Appl. Sci.* **2022**, *12*, 2928. [CrossRef]
39. Hu, H.; Cao, X.; Chen, Z.; Xie, J. Three-Dimensional Printing Materials for Cultural Innovation Products of Historical Buildings. *Buildings* **2022**, *12*, 624. [CrossRef]
40. Obradović, M.V.I.; Đurić, I.; Kićanović, J.; Stojaković, V.; Obradović, R. Virtual reality models based on photogrammetric surveys—A case study of the iconostasis of the serbian orthodox cathedral church of saint nicholas in Sremski Karlovci (Serbia). *Appl.Sci.* **2020**, *10*, 2743. [CrossRef]
41. Štroner, M.; Křemen, T.; Urban, R. Progressive Dilution of Point Clouds Considering the Local Relief for Creation and Storage of Digital Twins of Cultural Heritage. *Appl. Sci.* **2022**, *12*, 11540. [CrossRef]
42. Apollonio, F.I.; Fantini, F.; Garagnani, S.; Gaiani, M. A Photogrammetry-Based Workflow for the Accurate 3D Construction and Visualization of Museums Assets. *Remote Sens.* **2021**, *13*, 486. [CrossRef]
43. Marra, A.; Gerbino, S.; Greco, A.; Fabbrocino, G. Combining Integrated Informative System and Historical Digital Twin for Maintenance and Preservation of Artistic Assets. *Sensors* **2021**, *21*, 5956. [CrossRef] [PubMed]
44. Funari, M.F.; Hajjat, A.E.; Masciotta, M.G.; Oliveira, D.V.; Lourenço, P.B. A Parametric Scan-to-FEM Framework for the Digital Twin Generation of Historic Masonry Structures. *Sustainability* **2021**, *13*, 11088. [CrossRef]
45. De Marco, R.; Parrinello, S. Management of Mesh Features in 3D Reality-Based Polygonal Models to Support Non-Invasive Structural Diagnosis and Emergency Analysis in the Context of Earthquake Heritage in Italy. *Int. Arch. Photogramm. Remote Sens. Spat. Inf. Sci.* **2021**, *46*, 173–180. [CrossRef]
46. Gisbert Santaballa, A.G. La arqueología virtual como herramienta didáctica y motivadora. *Tecnol. Cienc. Y Educ.* **2019**, *13*, 119–147. [CrossRef]
47. Zhang, X.; Yang, D.; Yow, C.H.; Huang, L.; Wu, X.; Huang, X.; Guo, J.; Zhou, S.; Cai, Y. Metaverse for Cultural Heritages. *Electronics* **2022**, *11*, 3730. [CrossRef]
48. Selim, G.; Jamhawi, M.; Abdelmonem, M.G.; Ma'bdeh, S.; Holland, A. The Virtual Living Museum: Integrating the Multi-Layered Histories and Cultural Practices of Gadara's Archaeology in Umm Qais, Jordan. *Sustainability* **2022**, *14*, 6721. [CrossRef]
49. Skublewska-Paszkowska, M.; Powroznik, P.; Smolka, J.; Milosz, M.; Lukasik, E.; Mukhamedova, D.; Milosz, E. Methodology of 3D Scanning of Intangible Cultural Heritage—The Example of Lazgi Dance. *Appl. Sci.* **2021**, *11*, 11568. [CrossRef]

50. Youn, H.-C.; Yoon, J.-S.; Ryoo, S.-L. HBIM for the Characteristics of Korean Traditional Wooden Architecture: Bracket Set Modelling Based on 3D Scanning. *Buildings* **2021**, *11*, 506. [CrossRef]
51. Mohammadi, M.; Rashidi, M.; Mousavi, V.; Karami, A.; Yu, Y.; Samali, B. Quality Evaluation of Digital Twins Generated Based on UAV Photogrammetry and TLS: Bridge Case Study. *Remote Sens.* **2021**, *13*, 3499. [CrossRef]
52. Kęsik, J.; Miłosz, M.; Montusiewicz, J.; Israilova, N. Documenting Archaeological Petroglyph Sites with the Use of 3D Terrestrial Laser Scanners—A Case Study of Petroglyphs in Kyrgyzstan. *Appl. Sci.* **2022**, *12*, 10521. [CrossRef]
53. Poux, F.; Valembois, Q.; Mattes, C.; Kobbelt, L.; Billen, R. Initial User-Centered Design of a Virtual Reality Heritage System: Applications for Digital Tourism. *Remote Sens.* **2020**, *12*, 2583. [CrossRef]
54. Montusiewicz, J.; Barszcz, M.; Korga, S. Preparation of 3D Models of Cultural Heritage Objects to Be Recognised by Touch by the Blind—Case Studies. *Appl. Sci.* **2022**, *12*, 11910. [CrossRef]
55. Khalid, A. Conservation Challenges and Emerging Trends of Digital Preservation for UNESCO Architectural Heritage, Pakistan. *Conservation* **2021**, *2*, 26–37. [CrossRef]
56. Monego, M.; Fabris, M.; Menin, A.; Achilli, V. 3-D Survey Applied to Industrial Archaeology by Tls Methodology. *ISPRS—Int. Arch. Photogramm. Remote Sens. Spat. Inf. Sci.* **2017**, *42*, 449–455. [CrossRef]
57. Azmi, M.A.A.M.; Abbas, M.A.; Zainuddin, K.; Mustafar, M.A.; Zainal, M.Z.; Majid, Z.; Idris, K.M.; Ariff, M.F.M.; Luh, L.C.; Aspuri, A. 3D Data Fusion Using Unmanned Aerial Vehicle (UAV) Photogrammetry and Terrestrial Laser Scanner (TLS). In Proceedings of the Second International Conference on the Future of ASEAN (ICoFA), Singapore, 13–15 May 2017; pp. 295–305.
58. Intignano, M.; Biancardo, S.A.; Oreto, C.; Viscione, N.; Veropalumbo, R.; Russo, F.; Ausiello, G.; Dell'Acqua, G. A Scan-to-BIM Methodology Applied to Stone Pavements in Archaeological Sites. *Heritage* **2021**, *4*, 3032–3049. [CrossRef]
59. Rodriguez Sánchez, C. *Levantamiento Y Propuesta de Accesibilidad en El Jardín Histórico de la Arguijuela de Arriba (Cáceres)*; Universidad de Extremadura: Cáceres, Spain, 2021.
60. Xi, W.; Cong, W. Remote Practice Methods of Survey Education for HBIM in the Post-Pandemic Era: Case Study of Kuiwen Pavilion in the Temple of Confucius (Qufu, China). *Appl. Sci.* **2022**, *12*, 708. [CrossRef]

applied sciences

MDPI

Article

Hash Indexing-Based Image Matching for 3D Reconstruction

Mingwei Cao [1,2,*], Haiyan Jiang [1,2] and Haifeng Zhao [1,2]

[1] Anhui Provincial Key Laboratory of Multimodal Cognitive Computation, Anhui University, Hefei 230601, China
[2] School of Computer Science and Technology, Anhui University, Hefei 230601, China
* Correspondence: caomw@hfut.edu.cn

Featured Application: Our method can be used for 3D reconstruction, metaverse, virtual reality and augmented reality.

Abstract: Image matching is a basic task in three-dimensional reconstruction, which, in recent years, has attracted extensive attention in academic and industrial circles. However, when dealing with large-scale image datasets, these methods have low accuracy and slow speeds. To improve the effectiveness of modern image matching methods, this paper proposes an image matching method for 3D reconstruction. The proposed method can obtain high matching accuracy through hash index in a very short amount of time. The core of hash matching includes two parts: creating the hash table and hash index. The former is used to encode local feature descriptors into hash codes, and the latter is used to search candidates for query feature points. In addition, the proposed method is extremely robust to image scaling and transformation by using various verifications. A comprehensive experiment was carried out using several challenging datasets to evaluate the performance of hash matching. Experimental results show that the HashMatch presents excellent results compared to the state-of-the-art methods in both computational efficiency and matching accuracy.

Keywords: 3D reconstruction; image matching; structure from motion; local feature; feature matching

Citation: Cao, M.; Jiang, H.; Zhao, H. Hash Indexing-Based Image Matching for 3D Reconstruction. *Appl. Sci.* **2023**, *13*, 4518. https://doi.org/10.3390/app13074518

Academic Editors: Marek Milosz, Jacek Kęsik and Krzysztof Koszela

Received: 28 January 2023
Revised: 28 March 2023
Accepted: 31 March 2023
Published: 2 April 2023

1. Introduction

In recent years, image matching has attracted much attention from both computer vision and artificial intelligence communities and has been used in various practical applications, such as 3D reconstruction [1], augmented reality, virtual reality [2,3], visual localization, visual tracking, object detection, image classification [4,5], and medical image analysis. In the case of 3D reconstruction [6], the components of image-based 3D reconstruction mainly consist of image matching, camera pose estimation, triangulation, and bundle adjustment. Thus, image matching is one of the most important modules in image-based 3D reconstruction [7]. The goal of image matching is to generate feature correspondences across multiple views. Once feature correspondences have been generated, the structure from motion (SFM) system can be used to recover sparse point clouds and camera parameters, then the multi-view stereo and Poisson surface reconstruction can be utilized to generate dense point clouds and texture models, respectively. As a result, we can conclude that the quality of 3D reconstruction heavily depends on image matching methods.

The standard pipeline of image matching consists of feature point detection, feature descriptor computing, feature matching, and outlier removal [8]. Each component of image matching has many potential choices. In the case of feature point detection, there are scale-invariant feature transform (SIFT) [9], speeded-up robust features (SURF) [10], KAZE features [11], oriented fast and rotated brief (ORB) [12], and learned invariant feature transform (LIFT) [13]. These features can not only locate feature points but also generate feature descriptors. Moreover, there are also some deep learning-based methods, such as boosted efficient local image descriptor (BELID) [14], repeatable and reliable detector and descriptor (R2D2) [15], joint description and detection of local features (D2Net) [16], and SuperGlue [17]. According to reports in corresponding papers, every method has a state-of-the-art performance. For outlier removal, the ratio test [9] is the simplest but most efficient method, and it uses a constant threshold to reject incorrect feature correspondences and is widely used in various computer vision tasks. Ma et al. [18] proposed an automatic approach to remove outliers by vector field learning. This method has an excellent performance even when dealing with large-scale feature collections with heavy outliers.

For feature matching, there are also many methods. For example, brute force matching (BFM) [9] is one of the most famous methods. For a given query point, BFM chose a maximal confident candidate from reference points as the corresponding point. Usually, the feature correspondences generated by BFM have some outliers, often caused by the feature descriptor's ambiguity. Thus, BFM combined with the ratio test is a good solution for image matching. Wang et al. [19] proposed the use of a graph-matching approach to generate feature correspondences in the presence of outliers. The core idea is to reject incorrect matches by using the zero-assignment constraint. Jiang et al. [20] proposed the use of spatial clustering to generate a robust image matching result. The key idea of this method is to progressively cluster the putative matches into several motion-consistent clusters, including an outlier cluster. Zhou et al. [21] proposed progressive large-scale-invariant image matching in scale space for 3D reconstruction, resulting in a desirable performance in matching precision. Han et al. [22] proposed a deep learning-based approach to image matching as they designed a unified solution for learning feature representations and learning feature comparison functions. Although image matching has made significant progress, the topic still needs more attention, especially in image-based 3D reconstruction, and there is still much room for improvement in the accuracy and time efficiency of image matching methods.

After a deep investigation into the topic of image matching, we found that hash indexing is an efficient approach to feature descriptor matching. As a result, in this paper, we propose the design of a HashMatch based on hash indexing for image matching in 3D reconstruction. The HashMatch consists of two modules. First, constructing a big hash table for located feature points via Haar wavelet feature transform [23]. Second, indexing a potential candidate by using a short hash code for the query point. The HashMatch can significantly reduce computational costs on image matching when dealing with big image datasets. According to our experiments, the HashMatch has not only proved to be extremely fast but also has the desired matching precision and is very suitable for the processing of large-scale 3D reconstruction. In summary, the main contributions of this work are summarized as follows:

- We propose the use of hash indexing to replace the original brute force matching for image matching and present a HashMatch for 3D reconstruction.
- We propose the use of the Haar wavelet feature transformation to construct a big and robust hash table for the located feature points for feature descriptor indexing.
- We have designed a parallel architecture for hash matching to speed up the image matching process and save computing time. Additionally, a systematic experiment conducted on several benchmarking datasets is provided to assess the HashMatch against the state-of-the-art methods.

The rest of this paper is organized as follows: After a brief introduction in Section 1, related work about image matching is presented in Section 2. In Section 3, we present the HashMatch in detail. In Section 4, a systematic experiment was conducted on several benchmarking datasets to prove the effectiveness of the HashMatch. Finally, the conclusion and remarking comments are presented in Section 5.

2. Related Works

In this section, we will briefly review the most relative work to our research topic, including image matching, 3D reconstruction with SFM, MVS, and surface reconstruction.

2.1. Image Matching

Image matching has obtained great progress in the past decades due to many potential applications, so various feature-matching methods have been proposed by researchers [24,25]. In a nutshell, existing methods can be roughly divided into two categories according to the types of input data (video, and images). If the input data are in video form, the KLT and variants are usually adopted to locate feature correspondences by computing the displacements between two consecutive frames [26]. The Lucas and Kanade tracker (KLT) [27] is generally considered to be the best choice for video feature tracking, especially in robotic navigation, trajectory extraction for moving objects, and visual object tracking. Zach et al. [28] presented an adaptive KLT that was implemented in parallel programming for real-time applications. Hwangbo et al. [29] proposed the use of the inertial measurement unit (IMU) to improve the stability of KLT for a moving camera. Poling et al. [30] presented a novel framework for jointly tracking a collection of feature points, which can share information between the different feature points in the scene. This method achieves stable feature tracking via subspace constraints. Zhao et al. [31] formulated the feature tracking problem as a spatio-temporal statistical learning framework, so they proposed a robust keypoint tracking method via metric learning-driven multi-task structured output optimization. Generally, KLT and its variants can produce promising results for video sequences.

Regarding the second category, there have also been many approaches published in recent years. For example, Zhang et al. [31] proposed an efficient feature-tracking method to process a nonconsecutive image sequence for large-scale SFM, in which the parallel SIFT feature and second-pass matching strategy were adopted to handle feature-tracking drift. Lin et al. [32] developed a novel method called RepMatch for reconstructing the city model, in which the Epipolar guided approach is adopted to accommodate both wide baselines and repeated structures. Additionally, the RepMatch also utilized random sample consistency (RANSAC) to handle different true or false feature correspondences. Lukos et al. [33] proposed wide-baseline image matching via projective view synthesis for two-view image matching. This method can aptly handle the limitations caused by the approximate character of affine transformations via projective transformations, and can also use the essential matrix between images to reject incorrect feature correspondences. Recently, Lowry et al. [34] proposed to make use of local geometric priors to boost the spatial verification for filtering outliers. This method also utilized the scale and orientation information from feature neighborhoods to identify which feature points are likely to be correct feature correspondences and has proved to be excellent in practice.

Recently, the deep learning technique has provided excellent performances in various computer vision tasks, such as object detection and recognition, visual tracking and localization, and depth estimation. Thus, some researchers have tried to use deep learning techniques to handle image matching, including feature point detection, descriptor computing, and feature matching. For example, the first work of deep learning-based image matching was MatchNet, proposed by Han et al. [22], in which an end-to-end neural network architecture was designed to generate feature correspondences for image pairs. Zhao et al. [35] presented a hierarchical neural network named NM-Net to generate high-precision feature correspondences via mining reliable neighbors, in which NM-Net takes the generated graph as an input, then boosts the robustness to the order of feature correspondences. Yi et al. [36] designed a deep neural network architecture for the purpose of learning to locate good feature correspondences for wide-baseline images. They train the network in an end-to-end solution to label the feature correspondences as inliers or outliers. Sarlin et al. [17] proposed a neural network named SuperGlue to match two sets of local feature points by jointly finding feature correspondences and rejecting incorrect matches. Additionally, they also presented a flexible context aggregation mechanism by making use of visual attention to enable SuperGlue to reason the underlying geometry model. As a result, SuperGlue is an excellent solution based on deep learning for image matching. Ma et al. [37] presented a systematic survey to analyze the research state of image matching. More information about image matching can be found in Ma's work.

2.2. 3D Reconstruction

Reconstructing geometric models from image sets is a basic task in the fields of computer vision and computer graphics, which has led to the emergence of various 3D reconstruction methods. The classic pipeline of image-based 3D reconstruction consists of sparse, dense, and surface reconstruction. Each component has different solutions or methods. The goal of sparse reconstruction is to recover sparse point clouds and camera parameters from the generated feature correspondences via SFMs. For instance, Snavely et al. [38] developed a system named Bundler to recover sparse geometry from Internet image collections. Bundler possesses an excellent implementation and has the robustness to scene scales. Wu et al. [39] present a novel SFM system that reaches linear-time reconstruction via incremental way with the parallel SIFT and multi-core bundle adjustment. In addition to fast speeds, Wu's system can generate compact point clouds by re-triangulation for drifted features. Chris et al. [40] presented a large-scale SFM for outdoor scene reconstruction, in which they make use of the distributed camera model to describe image observations in terms of light rays rather than image pixels. Zhu et al. [41] developed a very large-scale global SFM to replace incremental SFMs, in which distributed motion averaging is utilized to estimate the camera pose in a parallel manner. Johannes et al. [42] revised the development progress of SFMs, then designed an excellent incremental SFM via scene graph augmentation. According to the reports in [43,44], modern SFMs can recover a world-scale model from Internet image collections in a few days.

For dense reconstruction, the patch-based multi-view stereo (PMVS) [45] is an excellent solution, which takes local feature points and edges as seeds to densify the sparse point clouds to generate dense point clouds via point expansion. Goesele et al. [46] proposed a depth-map-fusion-based MVS for generating dense point clouds. The method first computes individual depth maps using a stereo matching approach, then the depth maps are merged into a single dense model using a direct volumetric fusion. Silvano et al. [47] developed a massively parallel method for high-quality multi-view stereo matching, with the core idea being based on the PathMatch, which uses a robust photo-consistency measure to check depth errors. Kuhn et al. [48] proposed a scalable and efficient MVS method to reconstruct an accurate dense geometry model from hundreds of high-resolution images, in which all depth maps are estimated by semi-global matching (SGM) methods and a TV previously utilized to improve the quality of MVS. Inspired by the success of deep learning in pattern recognition, some researchers have tried to design neural networks for MVS reconstruction with end-to-end approaches. For example, Yao et al. [49] proposed the use of a deep learning-based approach to inference depth maps for generating dense point clouds. This resulted in the construction of a novel neural network architecture named MVSNet. Later, they modified MVSNet with recurrent neural networks (RNN), resulting in RMVSNet [50].

Once dense point clouds have been generated, surface reconstruction methods can be utilized to generate texture models. Among the existing methods, the most famous method is Poisson surface reconstruction [51], which formulates the problem of reconstructing the textured model as the Poisson equation solver. Zhou et al. [52] proposed the use of data-parallel Octrees to accelerate Poisson surface reconstruction when dealing with large-scale scenes, subsequently resulting in parallel Poisson surface reconstruction. Waechter et al. [53] presented a novel approach to generating high-quality textures for large-scale 3D reconstructions. This method can handle the complexity and scale of reconstructed models. Ummenhofer et al. [54] developed a variational method for surface reconstruction from the MVS data, which can handle a billion points in a short time. Recently, Tom et al. [55] presented a BigSUR for large-scale structured urban reconstruction, in which a binary integer program is proposed for global balances sources of error to generate semantically parsed mass models with associated façade elements. With the rapid development of deep learning technology, some learning-based approaches have been proposed for reconstructing texture models from image collections via end-to-end ways, such as SurfaceNet [56]. We believe that more efficient methods based on deep learning will be proposed in the future.

3. The Proposed Method

To improve the quality of modern 3D reconstruction, in this paper, we present an efficient image matching method based on hashing indexing, named HashMatch. The pipeline of the HashMatch is depicted in Figure 1, where we first use the SURF feature to find local feature points and compute feature descriptors for given image pairs. Second, Haar wavelet coding is adopted to construct a compact hash table for the located feature points and descriptors. Thirdly, the hash index replaces the traditional KNN to select the corresponding feature point from the reference feature set, and then the wrong feature correspondences are caused by repeating the query process. Note that the ratio test is also integrated into the HashMatch for rejecting outliers. The procedures of the HashMatch are presented in Algorithm 1. More details can be found in the subsequent sections.

Algorithm 1 HashMatch.

Input: Image collection $I = \{I_n | n \in [1, N]\}$
Output : Feature correspondences $M_{i,j} = \{f(p,q) | p,q \in [1, M]\}$

Step 1: Use the SURF feature to detect feature points for each image in image collection I.
Step 2: Use the SURF feature to compute feature descriptors for each located feature point.
Step 3: Create a hash table for all feature descriptors:
for idx = 1 to N':
(1) Compute Haar wavelet coefficients for each descriptor via Equation (8);
(2) Calculate the values of e_{k1}, e_{k2}, e_{k3}, and e_{k4} via Equations (9)–(12);
(3) Assign values to the members of hash table *htable* via Equations (13)–(16);
Step 4: Indexing hash table for given numbers i and j:
(1) Compute the values of id_0, id_1, and id_2 according to Equation (18);
(2) Calculate the indexing table for locating feature descriptors via Equation (19);
Step 5: Measure the difference between feature descriptor d_p and d_q via L2-distance;
Step 6: Repeat Step 4~Step 5. This should result in a set of feature correspondences for each image pairs.

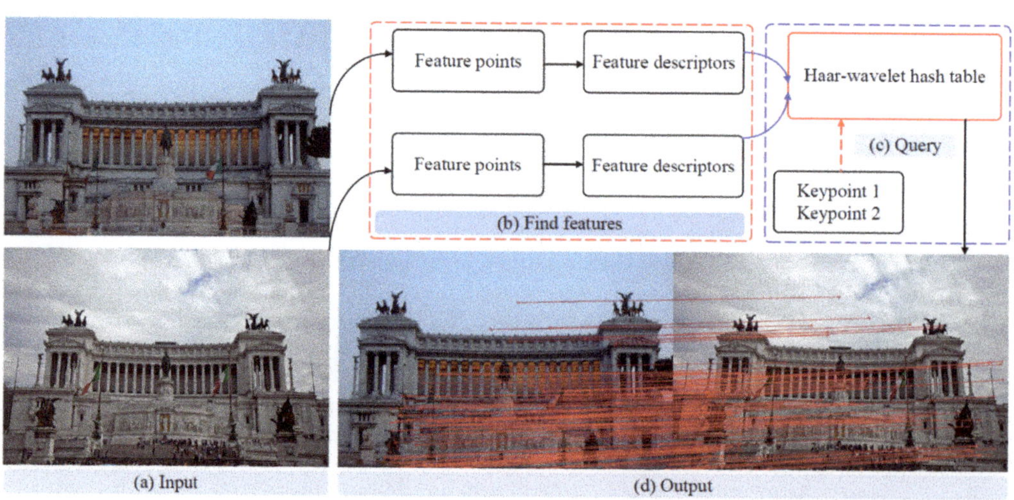

Figure 1. Flowchart of the HashMatch. The core of the proposed method consists of two parts: the first involves finding features including keypoint detection and computing feature descriptor, and the second involves matching feature descriptors by searching the created Haar wavelet hash table.

3.1. Feature Detection

Many feature point detectors can be candidates for image matching in 3D reconstruction, such as SIFT [9], ORB [12], KAZE [11], SuperPoint [57], and SuperGlue [17]. In this paper, the SURF [10] feature is adopted to find feature points and compute feature descriptors due to its fast speed and comparable matching precision to SIFT. Figure 2 presents the pipeline of the SURF detector for locating feature points from the given image. Specifically, for the input image I, the box filter is utilized to construct multi-scale spaces via different kernel sizes, and the integral image operation is also adopted to accelerate convolutional operation.

$$I_{\propto}(x,y) = \sum_{\substack{0 \leq x' < kernel.cols \\ 0 \leq y' < kernel.rows}} kernel(x',y') * I(x+x'-anchor.x, y+y'-anchor.y) \quad (1)$$

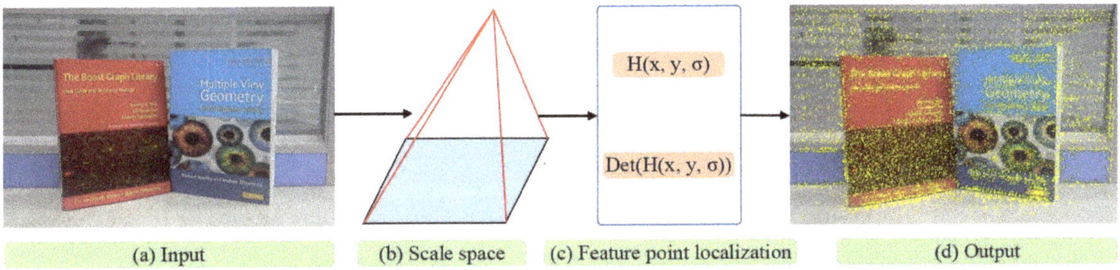

Figure 2. Flowchart of the SURF detector. The core idea of SURF consists of two steps: the first constructs a multiple-scale space for each input image, then the second locates the feature-point position.

The position of the feature point can be computed by the Hessian matrix in multi-scale spaces. The Hessian matrix $H(x,y,\sigma)$ in (x,y) at scale, σ is defined as

$$H(x,y,\sigma) = \begin{bmatrix} L_{xx}(x,y,\sigma) & L_{xy}(x,y,\sigma) \\ L_{xy}(x,y,\sigma) & L_{yy}(x,y,\sigma) \end{bmatrix} \quad (2)$$

where $L_{xx}(x,y,\sigma)$ is the convolution of the Gaussian second-order derivative $G(x,y,\sigma)$ with image I in point (x,y).

$$L_{xx}(x,y,\sigma) = \frac{\partial^2 g(x,y,\sigma)}{x^2} \otimes I(x,y,\sigma) \quad (3)$$

Notice that $L_{xy}(x,y,\sigma)$ and $L_{yy}(x,y,\sigma)$ can be calculated similarly.

To this end, we can judge whether (x,y) is a feature point according to the approximate determinant $\text{Det}(H'(x,y,\sigma))$ of the Hessian matrix $H(x,y,\sigma)$.

$$\text{Det}(H'(x,y,\sigma)) = L_{xx}(x,y,\sigma)L_{yy}(x,y,\sigma) - \left(0.9L_{xy}(x,y,\sigma)\right)^2 \quad (4)$$

If $\text{Det}(H'(x,y,\sigma))$ is greater than all pixel values in the neighborhood of (x,y), then (x,y) is a feature point, otherwise, it is not. As depicted in Figure 2, the circles containing yellow are the located feature points.

3.2. Descriptor Extraction

The descriptor is a representation of the local feature point, some novel descriptors have been proposed in recent years. For example, Lowe's SIFT and its variants, AKAZE feature [11], local difference binary (LDB) [58], locally uniform comparison image descriptor (LUCID) [59], learned arrangements of three patch codes (LATCH) [60], learned invariant feature transform (LIFT) [13], and SURF [10]. Binary descriptors possess a fast computational efficiency but have low matching precision. Floating-point descriptors have high matching precision but slow speed. There is an exception, namely SURF, which is a floating-point descriptor extractor, but it has comparable computational efficiency to binary descriptor extractors such as ORB and LDB. As a result, in this paper, we also use SURF to compute feature descriptors for located feature points due to it providing the best balance between distinctiveness and computational cost.

Figure 3 presents the pipeline for extracting feature descriptors for the located feature points via the SURF feature descriptor extractor. For the given feature point $\mathrm{x} = (x, y)$ and corresponding image I, we first compute a dominant orientation θ for feature point x by computing Haar wavelet responses in the neighborhood centered at x with radius $6s$, where $s = 1.2h/9$ and h is the size of the current filter kernel. With the dominant orientation θ, we crop the patch p with the size of $20s \times 20s$ centered at x from image I, then split the patch p into 16 sub-regions along with the dominant orientation θ, as depicted in Figure 3c, where each sub-region has four-types of Haar wavelet responses, namely

$$Resp_i = \left\{ \sum_i dx, \sum_i |dx|, \sum_i dy, \sum_i |dy| \right\} \tag{5}$$

where i denotes the index of sub-regions.

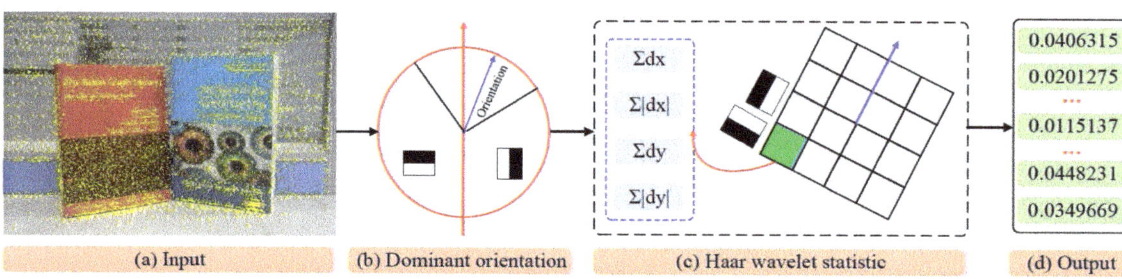

Figure 3. The pipeline of extracting feature descriptors.

To this end, we can obtain a floating-point feature vector with 64 elements.

$$D_x = \{Resp_i | i = 1, 2, \cdots, 16\} \tag{6}$$

Furthermore, we can normalize D_x by the Frobenius normal to boost its robustness to light and scale changes, and repeated texture.

3.3. Hash Matching

The goal of feature descriptor matching is to find feature correspondences for given independent feature collections. In the past decade, many approaches to feature descriptor matching have been proposed by researchers from academia. The simplest method is a k-nearest neighbor (KNN) matching with a ratio test. However, KNN is time-consuming, especially in large-scale feature collections. Johannes et al. [42] formulate feature descriptor matching as image searching for large-scale 3D reconstruction to reduce the computational cost of image matching, but matching precision may decrease in practice. Although there are some GPU-accelerated methods for image matching, they can also be improved in terms of both computational burden and matching precision. To improve the quality of image matching for boosting modern 3D reconstruction, in this paper, we propose a simple but efficient approach to feature descriptors matching via hashing indexing; thus, the proposed method is named hash indexing.

The pipeline of hash indexing is depicted in Figure 4, where feature descriptors are computed by the SURF feature [10]. For a given set of feature descriptors, the following steps should be conducted to generate feature correspondences: (1) create a hash table for all feature descriptors that included the initial hash table, feature transformation, and updated feature bins; (2) perform hash matching, specifically searching feature correspondences using the hash value of feature descriptors. In the implementation, we can make use of the Haar wavelet to create an approximate hash table for the acceleration of feature descriptors that match further. According to the report in Section 4, the proposed method outperforms the state-of-the-art methods in both efficiency and accuracy. It is salient to note that hash matching belongs to similar matching rather than brute-force matching; thus, it has a faster speed and desired matching precision.

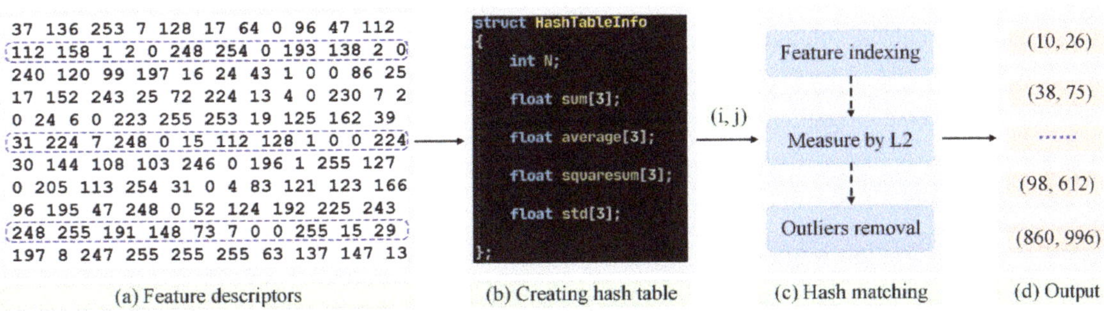

Figure 4. Flowchart of hash matching. Where (**a**) denotes the input feature descriptors, (**b**) represents the structure of the hash table in C-plus-plus language, (**c**) represents the pipeline of hash matching, and (**d**) denotes the output indexes of feature correspondences.

3.3.1. Creating a Hash Table

Descriptor matching is still a challenging task in large-scale 3D reconstruction. Although some novel approaches have been proposed for accelerating descriptor matching in image matching, these methods need further improvements regarding how they deal with high-resolution images. To reduce the computational costs of descriptor matching, in this section, we propose the creation of a big hash table to accelerate image matching via hash indexing, then boost the computational efficiency of image-based 3D reconstruction. Specifically, let *htable* be a hash table for the given descriptor collections $d = \{d_1, \cdots, d_N\}$, and the *htable* is defined as (in C-Plus-Plus language):

$$htable = \{int\ N; float\ sum[3]; float\ avg[3]; float\ squsum[3]; float\ std[3]\} \tag{7}$$

where N is the number of feature points, $sum[3]$ is used to store the sum of Haar wavelet coefficients of all feature descriptors and $squsum[3]$ is utilized to store the squared sum of Haar wavelet coefficients of all feature descriptors. Let c_{k1}, c_{k2}, and c_{k3} represent the Haar wavelet coefficients for the descriptor $d_k \in d$, which can then be calculated using the following formulas,

$$\begin{cases} c_{k1} = e_{k1} + e_{k2} - e_{k3} - e_{k4} \\ c_{k2} = -e_{k1} + e_{k2} - e_3 + e_{k4} \\ c_{k3} = e_{k1} - e_{k2} - e_{k3} + e_{k4} \end{cases} \tag{8}$$

Additionally, e_{k1}, e_{k2}, e_{k3}, and e_{k4} are the sum of descriptor elements and are defined as

$$e_{k1} = \sum_{i=0}^{i=3} \sum_{j=0}^{j=3} d_k[i * 8 + j] \tag{9}$$

$$e_{k2} = \sum_{i=0}^{i=3} \sum_{j=4}^{j=7} d_k[i * 8 + j] \tag{10}$$

$$e_{k3} = \sum_{i=4}^{i=7} \sum_{j=0}^{j=3} d_k[i * 8 + j] \tag{11}$$

$$e_{k4} = \sum_{i=4}^{i=7}\sum_{j=4}^{j=7} d_k[i*8+j] \tag{12}$$

Upon finding the values of c_{k1}, c_{k2}, and c_{k3}, the members of *htable* can be computed as

$$sum[index] = \sum_{k=0}^{k=N} c_{k1}, index = 0,1,2 \tag{13}$$

$$squsum[index] = \sum_{k=0}^{k=N} c_{k1} * c_{k1}, index = 0,1,2 \tag{14}$$

$$std[index] = \sqrt{\frac{squsum[index]}{N} - sum[index] * sum[index]}, index = 0,1,2 \tag{15}$$

$$avg[index] = \frac{sum[index]}{N}, index = 0,1,2 \tag{16}$$

To this end, we can create a big hash table based on Haar wavelet transformation for the input feature descriptor collections for fast indexing in the image matching stage.

3.3.2. Hash Indexing

The final stage of hash matching is to index hash table *htable* for the given feature descriptor $d_k \in d$. Brute force searching is the simplest method for indexing the hash table, but it may demand high computational costs. To mitigate this, we propose the construction of an index table *indtable* for the purpose of accelerating image matching. The *indtable* is a multi-dimensional array, which can be defined as (in C Plus Plus language) $indtable := std :: vector\langle std :: vector\langle std :: vector\langle int\rangle\rangle\rangle()$. Specifically, let inv_std_{index}=$1/std_{index}$, $index = 0,\ 1,\ 2,$ and id_0, id_1, and id_2 represent indexing numbers. t_0, t_1, and t_2 denote temporal variables, namely

$$\begin{cases} t_0 = indinv\_std_0 * (c_{k0} - avg[0]) \\ t_1 = indinv\_std_1 * (c_{k1} - avg[1]) \\ t_2 = indinv\_std_2 * (c_{k2} - avg[2]) \end{cases} \tag{17}$$

The values of id_0, id_1, and id_2 can be calculated by the following formulas.

$$\begin{cases} id_0 = \min(9, \max(0, floor(1.5 * (t_0 + 3.333)))) \\ id_1 = \min(9, \max(0, floor(1.5 * (t_1 + 3.333)))) \\ id_2 = \min(9, \max(0, floor(1.5 * (t_2 + 3.333)))) \end{cases} \tag{18}$$

where $\min(x, y)$ and $\max(x, y)$ represents the minimal and maximal values between x and y, respectively. To obtain an integer that is less than x, $floor(x)$ is used. Let $bins = std :: vector\langle std :: vector\langle int\rangle\rangle()$, and $index_t = 100 * id_0 + 10 * id_1 + id_2$. Thus, $bins[index_t] = k$, where k represents the indexing number of the kth feature descriptor in the image I_n, $n \in [0, N-1]$. As a result, the value of the indexing table can be assigned using the following formula.

$$indtable[n] = bins_n,\ n \in [0, N-1] \tag{19}$$

Once we have the indexing table, we can quickly obtain the descriptor collections for the image I_n and image I_{n+1} via hash indexing. The remaining operations of feature descriptor matching are similar to [23]. More detailed information which demonstrates the performance of the HashMatch can be found in Section 4, where a comprehensive experiment conducted on several challenging datasets is provided for readers.

4. Experimental Results

The HashMatch was implemented in Visual C++ 2017 with OpenCV SDK 4.1, and Nvidia CUDA SDK 10.0 version was also used to accelerate the process of image matching. Furthermore, we evaluated the HashMatch on the benchmarking dataset and the open-source dataset and also made a comprehensive comparison to KNN + RatioTest [9], ENFT [61], MODS [62], RepMatch [32], CODE [63], D2D [64], and R2D2 [15]. It should be noted that, during the experiment, each algorithm used the original parameters.

4.1. Evaluation on the EdgeFoci Dataset

The EdgeFoci dataset [65] is a benchmark for evaluating the performance of local feature detectors and descriptors. It consists of nine sequences and each sequence has a variant number of images. The samples of the EdgeFoci dataset are provided in Figure 5, they are Boat, Graffiti, Light, Dame, Obama, Painted Ladies, Rushmore, and Yosemite, respectively. The partially matching results are provided in Figure 6 where the BFM has the most incorrect matches. The RatioTest can reject some outliers from the matching collection produced by the BFM. The ENFT has more robustness than that of the BFM and KNN + RatioTest due to the use of a two-pass matching strategy. The HashMatch can generate the most accurate feature correspondences via Haar wavelet encoding. In the case of Notre Dame, it has symmetric structures and repeated features. The BFM produced the highest number of feature correspondences for Notre Dame, but there are too many incorrect matches in the collection of generated feature correspondences. The KNN + RatioTest also has outliers because its performance suffers from the constant threshold. If a big threshold is used by the KNN, it may produce more feature correspondences than that of the KNN with a small threshold but have more incorrect matches. Otherwise, The KNN with a small threshold can reject more incorrect feature correspondence; however, it can also reject some correct ones. Although the ENFT can filter many outliers from the collection of the initial feature correspondences by the two-pass matching strategy, a few incorrect matches also exist. However, the HashMatch has the best performance among the compared methods, providing the highest number of correct feature correspondences.

Figure 5. Image pairs from the EdgeFoci dataset.

In addition to the qualitative analysis, we also provided the statistical results (number of correct feature correspondences) for the compared methods. The results are listed in Table 1, where the higher the number of correct feature correspondences, the better the performance of the corresponding method. As the stated before, the KNN + RatioTest has the worst performance and the least amount of feature correspondences. The HashMatch has the highest number of feature correspondences (only inliers) on the whole dataset, demonstrating excellent performance. Finally, for each sequence, we recorded the computational time taken by the HashMatch. These times are depicted in Figure 7. The HashMatch can match ten thousand local features in a small amount of time. Moreover, we also provide the computational time for each method in Figure 8, where it can be seen that our proposed HashMatch has the fastest speed. To this end, both qualitative and quantitative results have proven the excellent performance of the HashMatch.

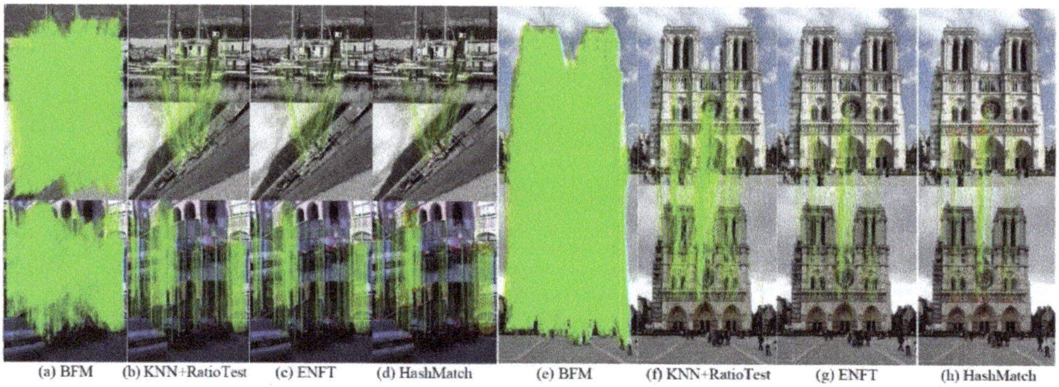

Figure 6. Matching results for Image pairs from the EdgeFoci dataset.

Table 1. Statistical results (correct matches) of the EdgeFoci dataset.

Method	Sequence							
	Boat	Graffiti	Light	Notre Dame	Obama	Painted Ladies	Rushmore	Yosemite
KNN + RatioTest	95	70	501	70	21	39	17	23
ENFT [61]	141	96	698	89	25	80	23	28
MODS [62]	101	83	564	142	42	47	18	32
RepMatch [32]	117	120	721	137	60	69	35	41
CODE [63]	98	109	706	110	58	53	29	47
D2D + KNN [64]	167	203	792	89	81	90	42	59
R2D2 + KNN [15]	130	178	807	120	90	83	53	62
HashMatch (Ours)	558	378	1263	307	328	407	286	613

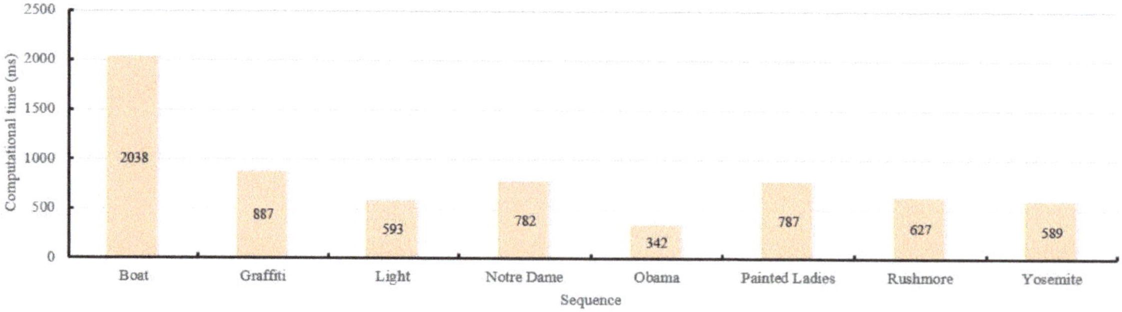

Figure 7. Computational times taken by the HashMatch on the Edge Foci dataset. These times consist of two parts, namely, feature detection time and feature matching time.

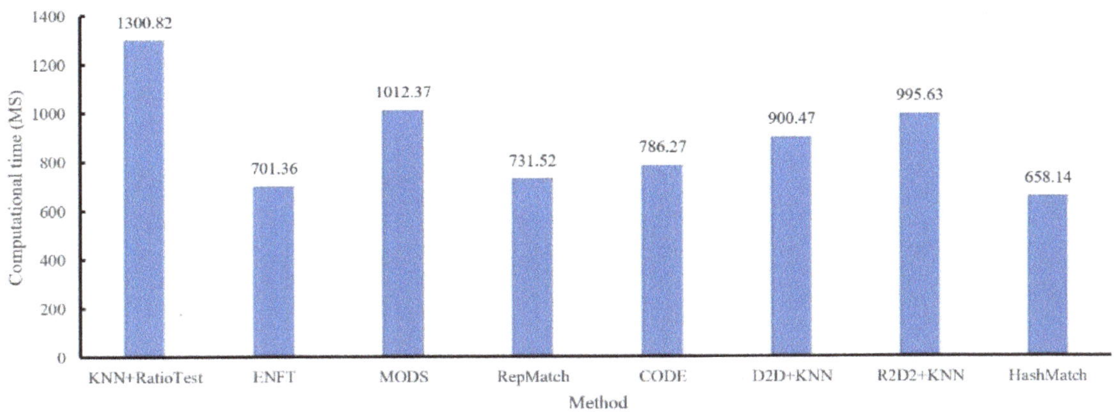

Figure 8. The calculation time of each method. According to these statistical results, the proposed method, HashMatch, has the fastest speed.

4.2. Evaluating HashMatch with Different Features

To further verify the effectiveness of the matching approach of the HashMatch, some different local features including BRISK, ORB, AKAZE, KAZE, SIFT, SURF & VGG, SIFT & VGG, have been integrated into the HashMatch, and we evaluated them using the Samper sequence of Strach's dataset (http://cvlabwww.epfl.ch/data/multiview/denseMVS.html, accessed on 10 April 2019). The statistical results are listed in Table 2, where the "Features" denote the number of input local features, and the "Matches" represent the number of generated feature correspondences. As shown in Table 2, "SURF + HM" has the highest number of feature correspondences; thus, the effectiveness of the HashMatch is proved once again. In addition to statistical results, we also provide some visualized results in Figure 9, in which all combiners can produce accurate feature correspondence. This is an excellent property of various computer vision tasks, especially in 3D reconstruction based on images.

Table 2. Statistical results (correct matches) of the Semper sequence of Strach's dataset, where HM represents HashMatch, "SURF + VGG" represents SURF detector and VGG descriptor, and "SIFT + VGG" represents SIFT detector and VGG descriptor.

Quantitative Analysis	Method							
	BRSK + HM	ORB + HM	AKAZE + HM	KAZE + HM	SIFT + HM	SURF + HM	SURF + VGG + HM	SIFT + VGG + HM
Features	19,808	20,000	11,896	11,434	38,991	45,944	45,944	38,991
Matches	3037	3857	4571	4699	6437	8539	7945	4808
Ratio of Inliers	79.34%	81.14%	88.56%	89.67	91.56	97.72	89.76	84.62

4.3. Evaluation on the Open-Source Dataset

To verify the effectiveness of the HashMatch in 3D reconstruction, we integrated it into the Bundler and evaluated it by using the open-source dataset (https://github.com/rperrot/ReconstructionDataSet, accessed on 12 April 2021). It should be noted that this dataset contains nine sequences. In our experiment, the "AvignonHotelDesMonnaies" (in the following section, we call it a "Hotel" for convenience) was adopted because it is the most challenging sequence. Figure 10 presents the samples of the Hotel sequence, there are many repeated features on the surfaces of the sampling images, and this raises a big challenge for the image matching method. The matching results of the compared methods are depicted in Figure 11, where it can be seen that the BFM has the most incorrect feature correspondences, meaning that BFM had the lowest matching precision among these image matching methods. Although the RatioTest can remove some outliers, there are also many incorrect feature correspondences in the collection of matches. The ENFT obtained a better result than that of the KNN + RatioTest because of its use of the two-pass matching strategy. The HashMatch yielded the best results among these methods as there were no outliers in the collection of feature correspondences.

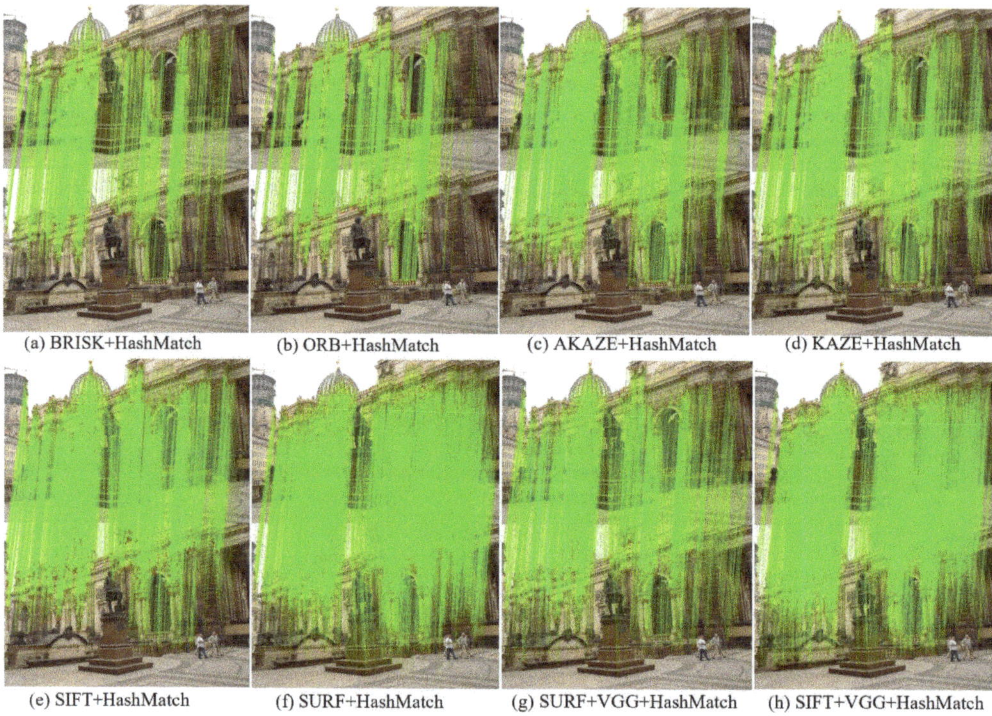

Figure 9. Feature matches of the Semper sequence, where the combination of "SURF + HashMatch" has the highest number of correct feature correspondences. Again, this proves the validity of the HashMatch method.

Figure 10. Samples of the Hotel sequence. Note that these figures have more repeated features and symmetrical structures.

Figure 11. Feature correspondences of the Hotel sequence. The HashMatch has the most correct feature correspondences. This proves that the HashMatch has more robustness to repeated features than that of the others.

Three ways are adopted to evaluate the performance of the HashMatch in 3D reconstruction, namely sparse reconstruction by Bundler, dense reconstruction by the PMVS, and surface reconstruction via the Poisson surface reconstruction. The sparse point clouds, dense point clouds, and textured models are provided in Figures 12–14, respectively. In each stage, the reconstructed model had a high geometry consistency with the real scene. Moreover, the HashMatch to one minute to generate feature correspondences. To this end, all of our results (Figures 11–14) prove the effectiveness of the HashMatch in 3D reconstruction.

4.4. Evaluation on Strecha's Dataset

To further assess the effectiveness of the HashMatch in 3D reconstruction, we integrated it into COLMAP and evaluated its effectiveness by using Strecha's dataset [66]. It should be noted that this dataset contains six sequences. In our experiment, the "Herz-Jesu-P25" is used because it is the most challenging sequence. Figure 15 presents the samples of Herz-Jesu-P25, and it can be observed that there are many repeated features on the surfaces of the sampling images, and this raises a big challenge for the image matching methods. The matching results of the HashMatch are depicted in Figure 16. Only correct feature correspondences are visible, indicating that the HashMatch had the highest matching precision compared to state-of-the-art methods. In summary, the proposed HashMatch has the best performance matching precision.

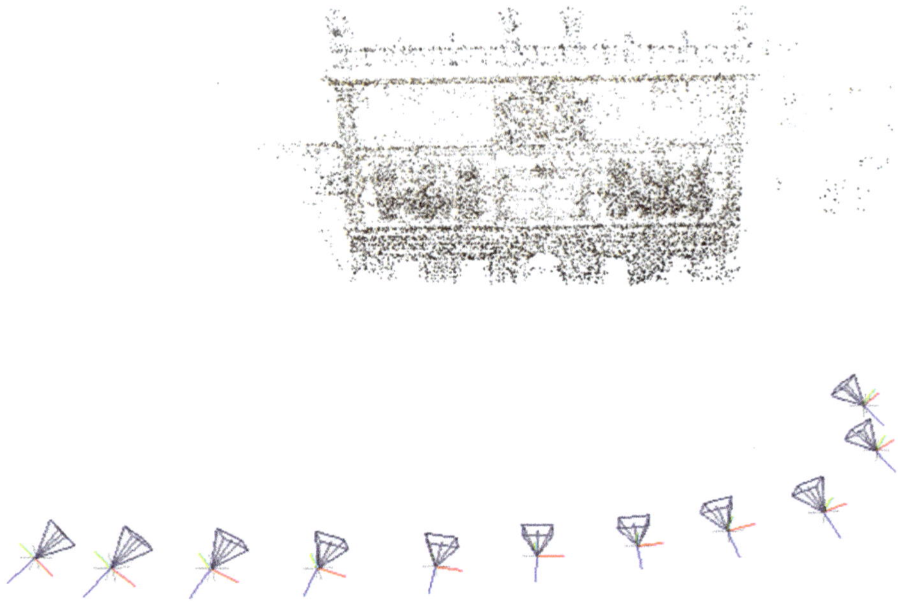

Figure 12. Sparse point clouds of the Hotel sequence, generated by Visual SfM with the HashMatch method.

Figure 13. MVS data of the Hotel sequence that was generated by the PMVS. The dense point clouds are very accurate. This is attributed to use of the HashMatch method.

Figure 14. Texture model of the Hotel sequence that was reconstructed by the Poisson surface reconstruction. The textured model has a high geometry consistency with the real scene.

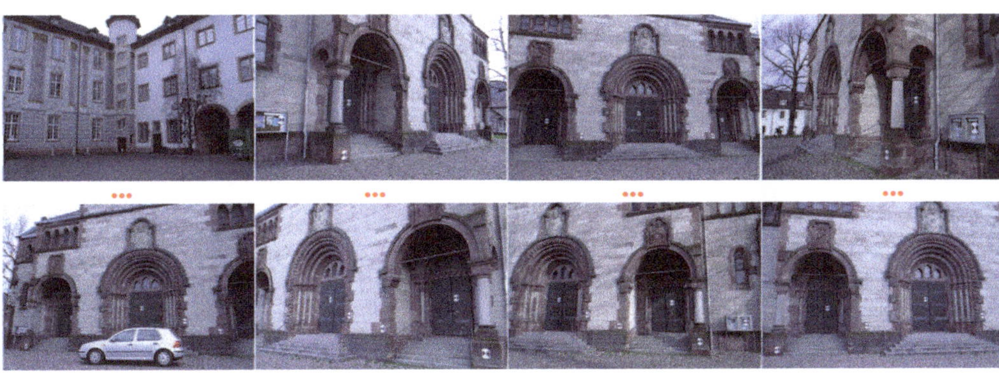

Figure 15. Samples of "Herz-Jesu-P25" in Strecha's dataset. These images have many repeated features. This poses a challenge to the feature matching method.

Figure 16. Feature correspondences of Herz-Jesu-P25 sequence in *Strecha's* dataset. Note that these feature correspondences are very accurate.

Three ways are used to evaluate the performance of the HashMatch in 3D reconstruction, namely sparse reconstruction by COLMAP, dense reconstruction by the PMVS, and surface reconstruction via the Poisson surface reconstruction. The sparse point clouds, dense point clouds, and textured models are provided in Figures 17–19, respectively, and in each stage, the reconstructed model has high geometry consistency with the real scene. Moreover, the HashMatch only took a few minutes (0.75 min) to generate feature correspondences. To this end, all of our results (Figures 16–19) prove the effectiveness of the HashMatch in 3D reconstruction.

Figure 17. Sparse point clouds of Herz-Jesu-P25 sequence, generated by Visual SfM with the HashMatch. These point clouds have accurate camera poses which are very important for dense reconstruction procedures.

Figure 18. Dense point clouds of Herz-Jesu-P25 sequence.

Figure 19. Texture model of Herz-Jesu-P25 sequence.

5. Conclusions

In this article, we propose an efficient image matching method for 3-D reconstruction. The proposed method was named HashMatch because it makes use of hash indexing to achieve large-scale feature matching in a short amount of time. We adopted the SURF feature to describe images, resulting in high-quality feature points and robust feature descriptors. We suggest using hash coding to generate a large hash table for the image set, so as to quickly index. Multiple verifications are utilized to reject incorrect feature correspondences. The results of systematic experiments show that the HashMatch is better than the state-of-the-art methods as it had the highest matching precision on all challenging datasets. In addition to this, the proposed method has good extensibility, which can work well with different feature detectors or feature descriptors. In short, our method is fast and accurate.

In the future, we also plan to design parallel architecture with the CUDA SDK for the HashMatch to provide more generic, fast, and accurate solutions for real-time image matching.

Author Contributions: Methodology, M.C.; supervision, H.Z.; funding acquisition, M.C., H.Z. and H.J.; Software, H.J.; investigation, H.J. All authors have read and agreed to the published version of the manuscript.

Funding: This work was supported in part by the National Natural Science Foundation of China (No. 61876002, No. 62076005, No. 62106006, No. U20A20398), Anhui Natural Science Foundation Anhui energy Internet joint fund (No. 2008085UD07), The University Synergy Innovation Program of Anhui Province, China (No. GXXT-2020-015, No. GXXT-2021-030, No. GXXT-2021-002, No. GXXT-2021-065), National Natural Science Foundation of China and Anhui Provincial Key Research and Development Project (No. 202104a07020029)).

Institutional Review Board Statement: Not applicable.

Informed Consent Statement: Not applicable.

Data Availability Statement: Data underlying the results presented in this paper are not publicly available at this time but may be obtained from the authors upon reasonable request.

Acknowledgments: We are grateful for the support and help of Guofeng Zhang in providing the ENFT code.

Conflicts of Interest: The authors declare no conflict of interest.

References

1. Zhu, R.; Guo, Z.; Zhang, X. Forest 3D Reconstruction and Individual Tree Parameter Extraction Combining Close-Range Photo Enhancement and Feature Matching. *Remote Sens.* **2021**, *13*, 1633. [CrossRef]
2. Wang, W.; Lv, Z.; Li, X.; Han, G.; Xu, W.; Zhang, B.; Zhu, Y.; Yan, Y. Spatial Query Based Virtual Reality GIS Analysis Platform. *Neurocomputing* **2018**, *274*, S0925231217306719. [CrossRef]
3. Lv, Z.; Li, X.; Li, W. Virtual reality geographical interactive scene semantics research for immersive geography learning. *Neurocomputing* **2017**, *254*, 71–78. [CrossRef]
4. Lan, R.; Lu, H.; Zhou, Y.; Liu, Z.; Luo, X. An LBP encoding scheme jointly using quaternionic representation and angular information. *Neural Comput. Appl.* **2019**, *32*, 4317–4323. [CrossRef]
5. Lan, R.; Zhou, Y.; Liu, Z.; Luo, X. Prior Knowledge-Based Probabilistic Collaborative Representation for Visual Recognition. *IEEE Trans. Cybern.* **2020**, *50*, 1498–1508. [CrossRef]
6. Cao, M.; Zheng, L.; Jia, W.; Lu, H.; Liu, X. Accurate 3-D Reconstruction Under IoT Environments and Its Applications to Augmented Reality. *IEEE Trans. Ind. Inform.* **2021**, *17*, 2090–2100. [CrossRef]
7. Cao, M.; Zheng, L.; Jia, W.; Liu, X. Joint 3D Reconstruction and Object Tracking for Traffic Video Analysis Under IoV Environment. *IEEE Trans. Intell. Transp. Syst.* **2020**, *22*, 3577–3591. [CrossRef]
8. Liang, A.; Li, Q.; Chen, Z.; Zhang, D.; Zhu, J.; Yu, J.; Fang, X. Spherically Optimized RANSAC Aided by an IMU for Fisheye Image Matching. *Remote Sens.* **2021**, *13*, 2017. [CrossRef]
9. Lowe, D.G. Distinctive image features from scale-invariant keypoints. *Int. J. Comput. Vis.* **2004**, *60*, 91–110. [CrossRef]
10. Bay, H.; Tuytelaars, T.; Van Gool, L. Surf: Speeded up robust features. In Proceedings of the European Conference on Computer Vision, Graz, Austria, 7–13 May 2006; Springer: Berlin/Heidelberg, Germany, 2006.
11. Alcantarilla, P.F.; Bartoli, A.; Davison, A.J. KAZE features. In Proceedings of the Computer Vision–ECCV, Florence, Italy, 7–13 October 2012; Springer: Berlin/Heidelberg, Germany, 2012; pp. 214–227.
12. Rublee, E.; Rabaud, V.; Konolige, K.; Bradski, G. ORB: An efficient alternative to SIFT or SURF. In Proceedings of the Computer Vision (ICCV) International Conference, Barcelona, Spain, 6–13 November 2011.
13. Yi, K.M.; Trulls, E.; Lepetit, V.; Fua, P. LIFT: Learned Invariant Feature Transform. *arXiv* **2016**, arXiv:1603.09114.
14. Suárez, I.; Sfeir, G.; Buenaposada, J.M.; Baumela, L. *BELID: Boosted Efficient Local Image Descriptor*; Springer International Publishing: Cham, Switzerland, 2019.
15. Revaud, J.; Weinzaepfel, P.; De Souza, C.; Pion, N.; Csurka, G.; Cabon, Y.; Humenberger, M. R2d2: Repeatable and reliable detector and descriptor. *arXiv* **2019**, arXiv:1906.06195.
16. Dusmanu, M.; Rocco, I.; Pajdla, T.; Pollefeys, M.; Sivic, J.; Torii, A.; Sattler, T. D2-Net: A Trainable CNN for Joint Description and Detection of Local Features. In Proceedings of the IEEE/CVF Conference on Computer Vision and Pattern Recognition (CVPR), Long Beach, CA, USA, 15–20 June 2019.
17. Sarlin, P.-E.; DeTone, D.; Malisiewicz, T.; Rabinovich, A. Superglue: Learning feature matching with graph neural networks. In Proceedings of the IEEE/CVF Conference on Computer Vision and Pattern Recognition, Seattle, WA, USA, 13–19 June 2020.
18. Zhao, J.; Ma, J.; Tian, J.; Ma, J.; Zhang, D. A robust method for vector field learning with application to mismatch removing. *CVPR 2011* **2011**, *31*, 2977–2984.
19. Wang, F.; Xue, N.; Yu, J.; Xia, G. Zero-Assignment Constraint for Graph Matching With Outliers. In Proceedings of the IEEE/CVF Conference on Computer Vision and Pattern Recognition (CVPR), Seattle, WA, USA, 13–19 June 2020.
20. Jiang, X.; Ma, J.; Jiang, J.; Guo, X. Robust Feature Matching Using Spatial Clustering With Heavy Outliers. *IEEE Trans. Image Process.* **2020**, *29*, 736–746. [CrossRef]
21. Zhou, L.; Zhu, S.; Shen, T.; Wang, J.; Fang, T.; Quan, L. Progressive Large Scale-Invariant Image Matching in Scale Space. In Proceedings of the International Conference on Computer Vision (ICCV), Venice, Italy, 22–29 October 2017.
22. Han, X.; Leung, T.; Jia, Y.; Sukthankar, R.; Berg, A.C. Matchnet: Unifying feature and metric learning for patch-based matching. In Proceedings of the IEEE Conference on Computer Vision and Pattern Recognition, Boston, MA, USA, 7–12 June 2015.
23. Brown, M.; Szeliski, R.; Winder, S. Multi-image matching using multi-scale oriented patches. In Proceedings of the Computer Society Conference on Computer Vision and Pattern Recognition (CVPR'05), San Diego, CA, USA, 20–26 June 2005.
24. Yu, Y.F.; Xu, G.; Huang, K.K.; Zhu, H.; Chen, L.; Wang, H. Dual Calibration Mechanism based L2,p-Norm for Graph Matching. *IEEE Trans. Circuits Syst. Video Technol.* **2020**, *31*, 2343–2358. [CrossRef]
25. Yu, Y.; Xu, G.; Jiang, M.; Zhu, H.; Dai, D.; Yan, H. Joint Transformation Learning via the L2,1-Norm Metric for Robust Graph Matching. *IEEE Trans. Cybern.* **2019**, *31*, 2343–2358. [CrossRef]
26. Lucas, B.D.; Kanade, T. An iterative image registration technique with an application to stereo vision. In Proceedings of the International Joint Conference on Artificial Intelligence, Vancouver, BC, Canada, 24–28 August 1981.
27. Gehrig, D.; Rebecq, H.; Gallego, G.; Scaramuzza, D. EKLT: Asynchronous Photometric Feature Tracking Using Events and Frames. *Int. J. Comput. Vis.* **2019**, *128*, 601–618. [CrossRef]
28. Zach, C.; Gallup, D.; Frahm, J.-M. Fast gain-adaptive KLT tracking on the GPU. In Proceedings of the Computer Society Conference on Computer Vision and Pattern Recognition Workshops, Anchorage, AK, USA, 23–28 June 2008.
29. Hwangbo, M.; Kim, J.-S.; Kanade, T. Inertial-aided KLT feature tracking for a moving camera. In Proceedings of the IEEE/RSJ International Conference on Intelligent Robots and Systems, St Louis, MS, USA, 11–15 October 2009.

30. Poling, B.; Lerman, G.; Szlam, A. Better feature tracking through subspace constraints. In Proceedings of the Conference on Computer Vision and Pattern Recognition, Columbus, OH, USA, 23–28 June 2014.
31. Zhao, L.; Li, X.; Xiao, J.; Wu, F.; Zhuang, Y. Metric learning driven multi-task structured output optimization for robust keypoint tracking. In Proceedings of the 29th {AAAI} Conference on Artificial Intelligence, Austin, TX, USA, 25–30 January 2015; pp. 3864–3870.
32. Lin, W.-Y.; Liu, S.; Jiang, N.; Do, M.N.; Tan, P.; Lu, J. RepMatch: Robust Feature Matching and Pose for Reconstructing Modern Cities. In *Computer Vidion—ECCV 2016, Proceedings of the 14th European Conference on Computer Vision, Amsterdam, The Netherlands, 11–14 October 2016*; Springer: Berlin/Heidelberg, Germany, 2016.
33. Roth, L.; Kuhn, A.; Mayer, H. Wide-Baseline Image Matching with Projective View Synthesis and Calibrated Geometric Verification. *PFG–J. Photogramm. Remote Sens. Geoinf. Sci.* **2017**, *85*, 85–95. [CrossRef]
34. Lowry, S.; Andreasson, H. LOGOS: Local Geometric Support for High-Outlier Spatial Verification. In Proceedings of the International Conference on Robotics and Automation (ICRA), Brisbane, Australia, 21–25 May 2018.
35. Zhao, C.; Cao, Z.; Li, C.; Li, X.; Yang, J. NM-Net: Mining reliable neighbors for robust feature correspondences. In Proceedings of the Conference on Computer Vision and Pattern Recognition, Long Beach, CA, USA, 15–20 June 2019.
36. Yi, K.M.; Trulls, E.; Ono, Y.; Lepetit, V.; Salzmann, M.; Fua, P. Learning to Find Good Correspondences. In Proceedings of the IEEE/CVF Conference on Computer Vision and Pattern Recognition, Salt Lake City, UT, USA, 18–23 June 2018.
37. Ma, J.; Jiang, X.; Fan, A.; Jiang, J.; Yan, J. Image Matching from Handcrafted to Deep Features: A Survey. *Int. J. Comput. Vis.* **2021**, *129*, 23–79. [CrossRef]
38. Snavely, N.; Seitz, S.M.; Szeliski, R. Photo tourism: Exploring photo collections in 3D. In *ACM Transactions on Graphics (TOG)*; ACM: New York, NY, USA, 2006; pp. 835–846.
39. Wu, C. Towards linear-time incremental structure from motion. In Proceedings of the International Conference on 3D Vision-3DV, Seattle, WA, USA, 29 June–1 July 2013.
40. Sweeney, C.; Fragoso, V.; Höllerer, T.; Turk, M. Large Scale SfM with the Distributed Camera Model. In Proceedings of the 4th International Conference on 3D Vision (3DV), Stanford, CA, USA, 25–28 October 2016.
41. Zhu, S.; Zhang, R.; Zhou, L.; Shen, T.; Fang, T.; Tan, P.; Quan, L. Very Large-Scale Global SfM by Distributed Motion Averaging. In Proceedings of the IEEE/CVF Conference on Computer Vision and Pattern Recognition, Salt Lake City, UT, USA, 18–23 June 2018.
42. Schönberger, J.L.; Frahm, J.-M. Structure-from-motion revisited. In Proceedings of the Conference on Computer Vision and Pattern Recognition (CVPR), Las Vegas, NV, USA, 27–30 June 2016.
43. Heinly, J.; Schönberger, J.L.; Dunn, E.; Frahm, J.M. Reconstructing the world* in six days. In Proceedings of the Conference on Computer Vision and Pattern Recognition (CVPR), Boston, MA, USA, 7–12 June 2015.
44. Agarwal, S.; Snavely, N.; Simon, I.; Seitz, S.M.; Szeliski, R. Building rome in a day. In Proceedings of the 12th International Conference on Computer Vision, Kyoto, Japan, 27 September–4 October 2009.
45. Furukawa, Y.; Ponce, J. Accurate, Dense, and Robust Multi-View Stereopsis. In Proceedings of the Computer Society Conference on Computer Vision and Pattern Recognition, Minneapolis, MN, USA, 18–23 June 2007.
46. Goesele, M.; Curless, B.; Seitz, S.M. Multi-View Stereo Revisited. In Proceedings of the Computer Society Conference on Computer Vision and Pattern Recognition, New York, NY, USA, 17–22 June 2006.
47. Galliani, S.; Lasinger, K.; Schindler, K. Massively Parallel Multiview Stereopsis by Surface Normal Diffusion. In Proceedings of the International Conference on Computer Vision, Las Condes, Chile, 11–18 December 2015.
48. Kuhn, A.; Hirschmüller, H.; Scharstein, D.; Mayer, H. A TV Prior for High-Quality Scalable Multi-View Stereo Reconstruction. *Int. J. Comput. Vis.* **2017**, *124*, 2–17. [CrossRef]
49. Yao, Y.; Luo, Z.; Li, S.; Fang, T.; Quan, L. Mvsnet: Depth inference for unstructured multi-view stereo. In Proceedings of the European Conference on Computer Vision (ECCV), Munich, Germany, 8–14 September 2018.
50. Yao, Y.; Luo, Z.; Li, S.; Shen, T.; Fang, T.; Quan, L. Recurrent mvsnet for high-resolution multi-view stereo depth inference. In Proceedings of the IEEE Conference on Computer Vision and Pattern Recognition, Long Beach, CA, USA, 16–17 June 2019.
51. Kazhdan, M.; Bolitho, M.; Hoppe, H. Poisson surface reconstruction. In Proceedings of the Eurographics Symposium on Geometry Processing, Sardinia, Italy, 26–28 June 2006.
52. Zhou, K.; Gong, M.; Huang, X.; Guo, B. Data-Parallel Octrees for Surface Reconstruction. *IEEE Trans. Vis. Comput. Graph.* **2011**, *17*, 669–681. [CrossRef]
53. Waechter, M.; Moehrle, N.; Goesele, M. Let There Be Color! Large-Scale Texturing of 3D Reconstructions. In *Computer Vision—ECCV 2014, Proseedings of the 13th European Conference, Zurich, Switzerland, 6–12 September 2014*; Springer: Berlin/Heidelberg, Germany, 2014; pp. 836–850.
54. Ummenhofer, B.; Brox, T. Global, Dense Multiscale Reconstruction for a Billion Points. *Int. J. Comput. Vis.* **2017**, *125*, 82–94. [CrossRef]
55. Kelly, T.; Femiani, J.; Wonka, P.; Mitra, N.J. BigSUR: Large-scale Structured Urban Reconstruction. *Acm Trans. Graph.* **2017**, *36*, 204. [CrossRef]
56. Ji, M.; Gall, J.; Zheng, H.; Liu, Y.; Fang, L. SurfaceNet: An End-to-End 3D Neural Network for Multiview Stereopsis. In Proceedings of the International Conference on Computer Vision (ICCV), Venice, Italy, 22–29 October 2017.
57. DeTone, D.; Malisiewicz, T.; Rabinovich, A. Superpoint: Self-supervised interest point detection and description. In Proceedings of the IEEE Conference on Computer Vision and Pattern Recognition, Salt Lake City, UT, USA, 18–23 June 2018.

58. Yang, X.; Cheng, K.-T. LDB: An ultra-fast feature for scalable augmented reality on mobile devices. In Proceedings of the International Symposium on Mixed and Augmented Reality (ISMAR) IEEE Conference, Atlanta, GA, USA, 5–8 November 2012.
59. Ziegler, A.; Christiansen, E.; Kriegman, D.; Belongie, S.J. Locally uniform comparison image descriptor. In Proceedings of the Advances in Neural Information Processing Systems, Lake Tahoe, NV, USA, 3–6 December 2012.
60. Levi, G.; Hassner, T. LATCH: Learned Arrangements of Three Patch Codes. *arXiv* **2015**, arXiv:1501.03719.
61. Zhang, G.; Liu, H.; Dong, Z.; Jia, J.; Wong, T.-T.; Bao, H. ENFT: Efficient Non-Consecutive Feature Tracking for Robust Structure-from-Motion. *arXiv* **2015**, arXiv:1510.08012.
62. Mishkin, D.; Matas, J.; Perdoch, M. MODS: Fast and robust method for two-view matching. *Comput. Vis. Image Underst.* **2015**, *141*, 81–93. [CrossRef]
63. Lin, W.Y.; Wang, F.; Cheng, M.M.; Yeung, S.K.; Torr, P.H.S.; Do, M.N.; Lu, J. CODE: Coherence Based Decision Boundaries for Feature Correspondence. *IEEE Trans. Pattern Anal. Mach. Intell.* **2018**, *40*, 34–47. [CrossRef] [PubMed]
64. Tian, Y.; Balntas, V.; Ng, T.; Barroso-Laguna, A.; Demiris, Y.; Mikolajczyk, K. D2D: Keypoint Extraction with Describe to Detect Approach. *arXiv* **2020**, arXiv:2005.13605.
65. Zitnick, C.L.; Ramnath, K. Edge foci interest points. In Proceedings of the International Conference on Computer Vision (ICCV), Barcelona, Spain, 6–13 November 2011.
66. Seitz, S.M.; Curless, B.; Diebel, J.; Scharstein, D.; Szeliski, R. A Comparison and Evaluation of Multi-View Stereo Reconstruction Algorithms. In Proceedings of the Computer Society Conference on Computer Vision and Pattern Recognition, New York, NY, USA, 17–22 June 2006.

applied sciences

MDPI

Article

Use of Cloud-Based Virtual Reality in Chinese Glove Puppetry to Preserve Intangible Cultural Heritage

Der-Lor Way [1,*] and Yu-Hsien Wei [2]

1 Department of New Media Art, Taipei National University of the Arts, Taipei 112, Taiwan
2 Department of Computer Science and Engineering, National Sun Yat-sen University, Kaohsiung 804, Taiwan; joewei0226@gmail.com
* Correspondence: adler@newmedia.tnua.edu.tw

Abstract: Chinese traditional glove puppetry is a folk art with a long history. It is worth inheriting and safeguarding this distinguished intangible cultural traditional art using virtual reality. With this background, this study integrates the digital resources of glove puppetry from the perspective of satisfying users' performance needs. In this study, a multi-user, cloud-based virtual reality glove puppetry system was developed that enhances the classic works of glove puppetry. Each user has a unique perception of the virtual environment and can interact remotely. The system involves human–computer and human–human interactions. This study also describes the design and control of glove puppets. The virtual reality system provides a unique entertainment experience to users of all ages. Through a questionnaire administered to 30 subjects after the user play, this study investigated the operation and experience of this system. According to the research findings, the proposed cloud-based VR system is not only easy to use, but also helps to preserve traditional intangible culture. Our research has high theoretical value and can help preserve traditional glove puppetry. Our cloud-based virtual reality system offers a new application for disseminating and preserving intangible cultural heritage.

Keywords: intangible cultural heritage; glove puppetry; virtual reality; human–computer interaction; cloud network

Citation: Way, D.-L.; Wei, Y.-H. Use of Cloud-Based Virtual Reality in Chinese Glove Puppetry to Preserve Intangible Cultural Heritage. *Appl. Sci.* **2023**, *13*, 5699. https://doi.org/10.3390/app13095699

Academic Editors: Marek Milosz and Jacek Kęsik

Received: 18 March 2023
Revised: 2 May 2023
Accepted: 3 May 2023
Published: 5 May 2023

1. Introduction

Forms of intangible cultural heritage include practices, knowledge, skills, representations, objects, artifacts, instruments, communities, and cultural spaces associated therewith. Intangible cultural heritage is passed down through generations, constantly reinvented by communities in response to changes in their environment and history. Intangible cultural heritage affords communities with a sense of identity and stability and is a key aspect of cultural diversity and human creativity [1]. The "Convention for the Safeguarding" of the Intangible Cultural Heritage treaty proposed by the UNESCO (United Nations Educational, Scientific, and Cultural Organization) divides intangible cultural heritage into (1) oral traditions; (2) performance art; (3) social customs, etiquette, and festivals; (4) knowledge and practices about the environment; and (5) traditional crafts.

Intangible cultural heritage is more challenging to safeguard than tangible cultural heritage because performance art, music, and crafts have no distinct form. The conservation of intangible heritage is also complicated by its long history [2]. In addition, the COVID-19 pandemic in 2020 created new challenges for exhibition centers and apprenticeship training and prevented festivals about intangible heritage from being held, necessitating the creation of new methods of preservation and civic tourism [3]. However, the COVID-19 pandemic also created challenges for museums. Because museums were forced to close during the pandemic, they began experimenting with new methods of operating, such as online programs [4,5]. The online use of the museum not only created digital exhibitions, but also expanded the exhibition space. Virtual reality (VR) and augmented reality (AR) can enhance museum visitors' experiences [6]. Museums are exploring the use of digital content for exhibitions and experiences by conducting preference studies [7]. Considering these features, the use of VR and AR has increased in museum exhibitions involving intangible heritage.

In the digital age, which pursues sophistication and speed, traditional glove puppetry struggles to attract the attention of people. To inherit the legacy of traditional culture, many declining traditional skills have brought the younger generation's understanding of our country's cultural history through emerging technology and application. VR is an emerging technology that facilitates effective learning, communication, and entertainment within cultural spaces. The virtual displaying of cultural artifacts may involve interactive storytelling. Captivating storylines within virtually reconstructed three-dimensional (3D) environments can engage consumers and provide both educational and entertainment value. AR and VR can play key roles in presenting past and cultural practices through innovative methods [8]. Physical movement in VR can be an effective learning activity because it can be more motivating and engaging than traditional teaching methods [9,10]. This means that virtual reality can provide a superior understanding of the content of intangible cultural heritage.

The purpose of this study was to use cloud-based virtual reality in Chinese glove puppetry to preserve intangible cultural heritage. Therefore, this study developed a multi-user, cloud-based virtual reality glove puppetry system that enhances the classic works of glove puppetry, as shown in Figure 1. Each user has a unique perception of the virtual environment and can interact remotely. In addition, the Internet enables users to collaborate in different locations to perform together. To ensure that collaboration in VR is effective, participants' gestures and emotions must be captured and transmitted in real-time [11]. Therefore, teamwork allows players to work collaboratively in different places. Finally, a user experience study is included in this article. User experience research focuses on developing a system usability scale and provides an understanding of puppet cultural preservation. The main contributions of this study are as follows:

1. This paper presents a collaborative work of a puppetry show in a multi-user virtual reality system.
2. The user experience evaluation was implemented to identify predictive and interpretive use measures for these technologies.
3. The proposed cloud-based VR system is easy to use and is not limited to users of a particular group.
4. Our VR system not only provides experience of the operation puppetry but also helps preserve traditional intangible culture.

Figure 1. Chinese glove puppetry in cloud-based VR.

The remainder of this paper is arranged as follows. Section 2 reviews literature research. Section 3 describes glove puppetry digital design including stage and puppets. Section 4 presents the proposed cloud-based virtual reality system. A user experience study was implemented and is discussed in Section 5. Finally, Section 6 concludes the paper and provides suggestions for future work.

2. Literature Research

2.1. Chinese Glove Puppetry

Chinese glove puppetry incorporates performing art, visual art, music, classical literature, and folklore. The UNESCO included Chinese glove puppetry in its list of intangible cultural heritage in "Convention for the Safeguarding" in 2012 [12]. Glove puppetry is a type of performance in which puppets animate and create characters, stories, and worlds. Glove puppetry originated in China's Fujian province in the 17th century. Glove puppetry is not only an integral part of folk art, but also a long history of traditional art. Glove puppetry began as street entertainment and was introduced to Taiwan in the mid-19th century [13]. Glove puppetry has since become a popular art form in Taiwan and is often performed during celebrations, such as weddings and religious ceremonies [14]. Glove puppets have deeply influenced the daily life of agricultural societies. They often spread morality and social order, serving as beliefs in the Confucian spirit through the storytelling of a glove puppet show. In the 1950s, Golden Ray's puppets revolutionized the art form. The introduction of television in the 1960s enabled the art forms to develop further. Puppet shows were broadcast on Taiwanese television, and their popularity skyrocketed in the proceeding decade [15]. With the transition from an agricultural society to an industrial society, traditional glove puppetry face challenges such as shortages of trainees, audiences, and funds to preserve their heritage.

Generally, there are six typical roles: male protagonists, female protagonists, painted faces, elders, clowns, and monsters. Each character's makeup, facial expressions, clothing, props, and gestures express its characteristics [16], as shown in Figure 2. For example, a male protagonist's exquisite facial features and graceful clothes show that he is a garbed young male scholar. A painted face displays his confident, courageous, and adventurous warrior character. In traditional glove puppets, such archetypes and attributes are crucial because they allow the audience to quickly identify each puppet's role in the play. Traditionally, glove puppets have heads, hands, and feet made of wood. The puppeteer carefully carves a piece of wood into the shape of a hollow human head, which is then refined and polished. The most embroidered dresses and beautiful props are cloth decorated with absorbing tassels, beads, treasures, feathers, etc. The beautiful craftsmanship of glove puppets often attracts audiences.

Figure 2. Six typical character roles in puppetry Glove puppets [16].

Puppeteers must be highly skilled in manipulating puppets using their fingers. A puppeteer inserts his index finger into the hollow center of the puppet's head; his thumb controls one hand of the puppet; and the other three fingers hold the other hand of the puppet. Puppeteers bring puppets to life by skillfully manipulating their movements and gestures, as shown in Figure 3. Glove puppets have hollow wooden heads that are carved in the shape of a human head. The torso and limbs of the puppets are made of cloth. A hand enters the glove puppet costume and performs with it during the show. Moreover, glove puppetry performances involve sets and backcourt music. The scripts are passed down orally by storytellers and are typically adapted from classic novels such as Journey to the West, Romance of the Three Kingdoms, Water Margin, and The Seal of the Gods. Unfortunately, only a few traditional glove puppet shows have survived [13].

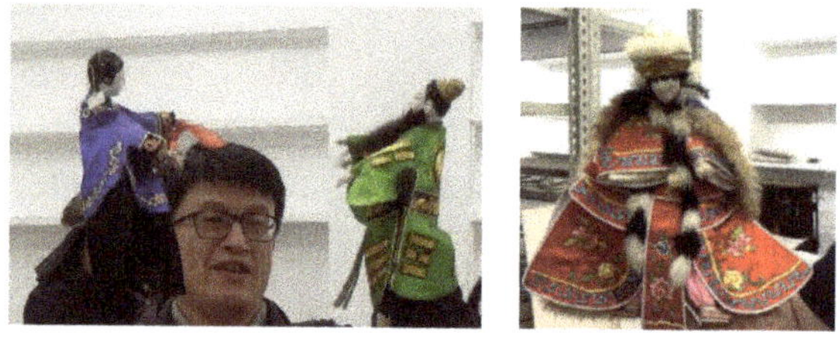

Figure 3. Glove puppets.

2.2. *Virtual Reality and Cultural Heritage*

Recently, proponents of cultural heritage are increasingly using VR in museum exhibitions. These technologies can create user-centric experiences and produce cultural heritage digitally accessible, especially when physical admission is constrained. Various methods of using VR to protect intangible cultural heritage have been proposed. Skublewska-Paszkowska et al. gathered many papers about 3D technologies for intangible cultural heritage preservation [17]. They investigated the use of 3D technologies to maintain intangible cultural heritage and discovered that intangible cultural heritage is a necessary extension of tangible cultural heritage. The growing attentiveness of intangible cultural heritage in scientific studies is noticeable. Societies have come to understand the cost of losing their intangible cultural heritage. Research on the use of 3D technologies such as 3D scanning, 3D modeling, VR, and AR to reproduce cultural heritage has increased. Zhao proposed a digital method for preserving intangible cultural heritage [2], including augmented reality technology. They examined the main characteristics of intangible cultural heritage, as well as the artifacts and practices of a culture that have been preserved and transmitted throughout generations. Their AR can greatly improve user experience.

Further, Day and Way established a VR program for a white crane folk dance performed in Mazu temples in Taiwan [18], as shown in Figure 4. In the first phase, the Kinect module was used to capture the performance of the folk dancers and animate the motion of the 3D animation characters (white crane). In the second phase, a Kinect and HTC (High Tech Computer Corporation) VIVE tracker were used to capture user actions. Using the Unity integration platform, the 3D animation characters played, and the 3D avatar to be instantly controlled was presented to the user in the environment. A VR headset was paired with the Kinect motion capture sensor to immerse users into a 3D virtual world where they could view the dancing cranes and appreciate a dynamic virtual environment in real time.

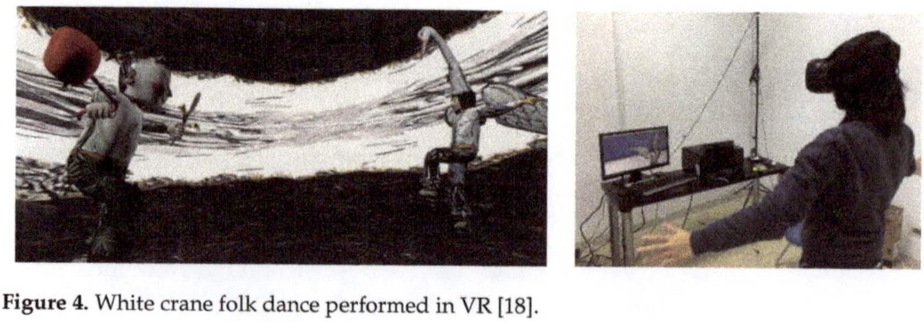

Figure 4. White crane folk dance performed in VR [18].

Selmanovic et al. improved user-friendliness in relation to intangible cultural heritage protection using virtual reality [19]. They investigated whether a VR-based 360-degree video could facilitate immersion in applications for the safeguarding of intangible cultural heritage and thereby increase its accessibility. Huang et al. provided the self-issuance method to the folk intangible cultural heritage center [20]. They presented a method of presenting intangible cultural artifacts by using VR headsets. Unlike traditional AR techniques, this approach enables users easily and fully to interact with intangible cultural heritage in an immersive virtual environment. Han and Yi investigated how digital exhibitions of intangible heritage were used during COVID-19 to help audiences understand intangible heritage [21]. They also explored how museums can use digital technology to reproduce intangible heritage for in-person experiences.

Nitsche and McBride explored controlling puppets in VR and puppet-based VR games [22]. Through their bottom-up design approach, they created a VR game in which users adopt the puppets perspective, thereby demonstrating the various design possibilities that VR offers. Lin et al. proposed a method for controlling a virtual puppet by using a computer [23]. Users can select from different puppets, scenes, and music. They used Leap motion controllers to capture the user's gestures and display the puppet's movements on a computer screen. Unfortunately, they only build a 3D scene with a fixed viewpoint. In fact, their system is a 3D multimedia system without immersion. So far, some above virtual reality systems for a single user have been established for intangible cultural heritage. No multi-user, cloud-based VR systems have been developed to present intangible cultural heritage. For this reason, this paper proposed a multi-user, cloud-based VR glove puppetry system to protect intangible cultural heritage.

2.3. *Virtual Reality Collaboration Systems*

VR collaboration systems are becoming increasingly popular for their ability to enhance collaboration through multi-user functionality. Interest and research in this technology have increased, with organizations seeking to capitalize on this. Researchers have explored aspects of remote collaboration that can facilitate communication and expression. Thomas proposed a low-cost, lightweight, easily configurable MuVR device (multi-user virtual reality) [24]. Their MuVR system is portable and without the requirement of a high-priced external infrastructure. Nguyen et al. established a CollaVR [25]: an application that enables multiple users to collaborate and review a VR video together through a network. Ibayashi et al. created a Dollhouse VR [26], a multi-user VR workspace for collaborative design. Piumsomboonet et al. delivered a multi-user VR and AR platform for remote collaboration [27]. Sra et al. provided a shared VR experience for remote users known as "Your Place and My Place" [28].

Some of the above-mentioned VR systems have been held back by poor accessibility and lack intuitive multi-user capabilities. MuVR is portable but has a low resolution. CollaVR is a video-based VR that is realistic but lacks interaction. Although Dollhouse VR and CoVAR allow users to collaborate on spatial tasks in a shared space, they are not suitable for cultural heritage applications. In fact, the system design and requirements are different depending on the application. Most VR applications require trained people to perform well in a virtual reality environment. Therefore, it is important that user experience studies evaluate system usability and requirements. Unfortunately, most of the aforementioned systems do not conduct user experience studies.

In summary, multi-user virtual reality is an emerging technology and a great tool for facilitating collaboration. For collaborative tasks in virtual reality systems, the gestures and sensations of immersed users must be accurately exchanged and visualized in real time. By using a VR periphery network, we developed a multi-user, cloud-based VR glove puppetry system in which users in different locations could perform together. Finally, our user study shows that our VR system is not only easy to use, but also helps preserve the traditional glove puppetry culture.

3. Glove Puppetry Design

As mentioned before, a puppet's head is a hollow carved wood with a human-like head. The puppet's body is composed of draped fabric clothes. A hand goes into the glove puppet costume for a performance. Moreover, the stage for the puppetry is the frontcourt in the puppet show. Over time, the design of glove puppetry stages has changed along with the values and customs of society. Puppet theaters were typically found in temples, town squares, and marketplaces. There was no fixed stage in the early period of puppetry. Shows could be staged on a simple and straightforward stage. A portable puppet stage known as the "four-corner shed" was invented in the Qing Dynasty. Later on, the stage design was modified into a colorful hexagonal shed.

3.1. Stage Design

Our digital stage was inspired by the stage of the Seden Puppet Theater Foundation, located in the library of Taipei National University of the Arts [29] (Figure 5a,b). This hexagonal shed is adorned with ornate gold leaf and intricately painted maps, creating resemblance to a temple. The stage consists of three parts: the upper cover, base, and screen column. The upper cover acts as the roof and is supported by dragon-shaped pillars, and the base contains window frames. The stage is divided into two floors, with three doors and three windows on each floor. A plaque is inlaid into the upper floor, featuring the name of the school elegantly engraved. Curtains are hung across each door, and the words "General" and "Chancellor" are displayed on the lower sections of the curtains. The base is as tall as the puppeteer's chest. A wooden frame beneath the base provides structural support. According to the above breakdown, Figure 6 displays our 3D stage design for a cloud-based VR puppet show.

(a) (b)

Figure 5. Seden puppet theater. (**a**) Seden puppet theatre. (**b**) Details of Seden puppet theater.

(a) (b)

Figure 6. Stage design for cloud-based VR puppet show. (**a**) Our stage design. (**b**) Details of our stage design.

3.2. Glove Puppet Design

Categories of glove puppet characters include the male protagonist, female protagonist, painted face, clown, child, beast, and other. Heroic characters tend to have striking features, with high foreheads and a majestic appearance. Villains tend to have bushy eyebrows, deep-set eyes, pronounced cheekbones, and large mouths. Figure 7 presents a male protagonist and a painted face. Male protagonists are often handsome and enchanting characters. The facial expressions and colors of the characters indicate personality traits, such as loyalty, treachery, kindness, clown, evil, etc.

(a) (b)

Figure 7. Glove puppet design. (**a**) Male protagonist. (**b**) Painted faces.

3.3. Clothing Simulation

The personality traits of the characters are also expressed through their different costumes. Simulating the dynamic movements of clothing to enhance the motion of a glove puppet is essential for an effective puppet show. The puppet is controlled by each of the fingers. The puppeteer's gestures are captured using a Leap Motion controller and transferred to the puppet's head and hands. Puppet costumes have three components: the sleeves, the front of the skirt, and the back of the skirt. The clothes of the glove puppet are held up by the fingers, and the texture inside is enhanced.

The dynamic joints were used to simulate realistic movement. The structure of the skeleton and the dynamic joint chains are presented in Figure 8. The hands are controlled by dynamic joint chains. The joints allow for rotation and translation between connected, rigid bodies. Different joints, such as rotary joints, ball joints, online point joints, and pulley joints, can be used to combine rigid bodies into a complete puppet. By defining the degrees of freedom of each joint, the puppet's motion can be constrained. The front and back of the skirt are controlled by five dynamic joint chains each, allowing the costume to swing up and down as the puppeteer moves their hands.

Figure 8. Structure of skeleton and dynamic joint chain of glove puppet.

4. Cloud-Based Virtual Reality System

This study generates a collaborative, cloud-based virtual reality system for glove puppetry, which is projected to perform a real-time interactive show for safeguarding intangible cultural heritage. The cloud-based VR system we developed is based on Unity. A Leap Motion controller is connected to an HTC Vive head-mounted display, and 3D data are imported into the system. We also improved core technologies and incorporated multi-user functionality into the system. Puppeteers can see each other and the movement of their characters to imitate the interactions with the virtual object.

The schema of virtual reality glove puppetry system is illustrated in Figure 9. When a user registers as a host, the server adds the user and creates a left-hand or right-hand puppet avatar in the scene. The user can control the puppet with their hand through separately acquired data. The Leap Motion controller connected to the HTC Vive VR headset captures the motion of the user's hand. The motion data are transmitted to the computer through the VRPN (VR periphery network). Figure 10 illustrates a user's index finger controlling the puppet's head; the user's thumb and little finger control the puppet's right and left hands, respectively. Two players can join in person, and additional players can join remotely over the Internet. The VR environment is displayed in real time on the connected computer monitor. So, each player can meet and see scenes performed synchronously on their own computer.

Figure 9. Schema of virtual reality glove puppetry system.

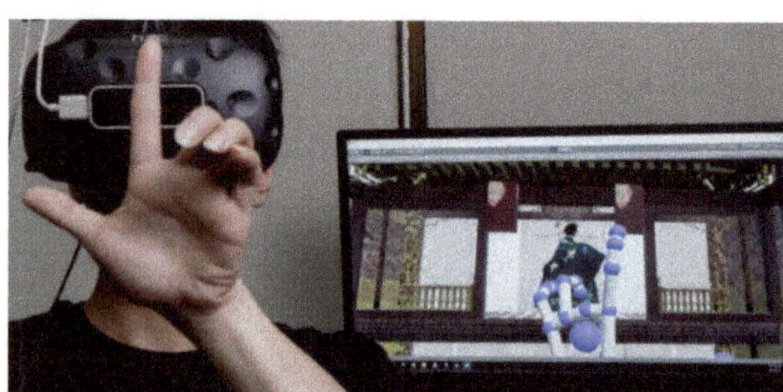

Figure 10. Puppeteer wearing HTC head-mounted display and controlling glove puppet avatar.

Data in our system are obtained from Leap Motion controllers and transmitted through the VR peripheral network. The system architecture of the VR glove puppetry system is presented in Figure 11. The network conversion component handles data conversion across the network. The process flow is as follows:

1. Users A and B: Users log into the system. Physical user data are obtained from sensors. For instance, the HTC Vive VR headset catches data on the user's situation and direction, and the Leap Motion controller obtains data on the user's hand. The Leap Rig transfers the data between the sensor and Unity.

2. VR periphery network registration: The Unity network architecture is used to create a client–server network for the VR periphery network. A network manager is spawned to handle network events.
3. Puppet avatars: Three tasks ("Network Identity," "Enable Ctrl," and "CCDIK") are generated during the prefabrication of a glove puppet avatar. Network Identity assigns each avatar a unique ID. Enable Ctrl adds "Avatar Controller" to enable animation when the native player starts. CCDIK (Cyclic Coordinate Descent Inverse Kinematics) handles inverse kinematics for custom skeletons.
4. Avatar controller and dynamic joint: The glove puppet avatar's head and hands are controlled by the user's thumb and index and little fingers, respectively. The avatar controller transfers captured user motion to the avatar. Dynamic joints are attached to the hand bones on the avatar's clothing to make the simulation more realistic.
5. Network transform and network skeleton: In the Unity network architecture, the client cannot communicate directly. Synchronization can only be achieved by the server. The network transform synchronizes the movement and orientation of a player's hands and head. The network skeleton synchronizes the skeleton data of each puppet.
6. Data synchronization: Each client transmits data to the server, and the server transmits data to other clients synchronously.
7. Puppet cloud theater: Puppet performances with users in different locations.

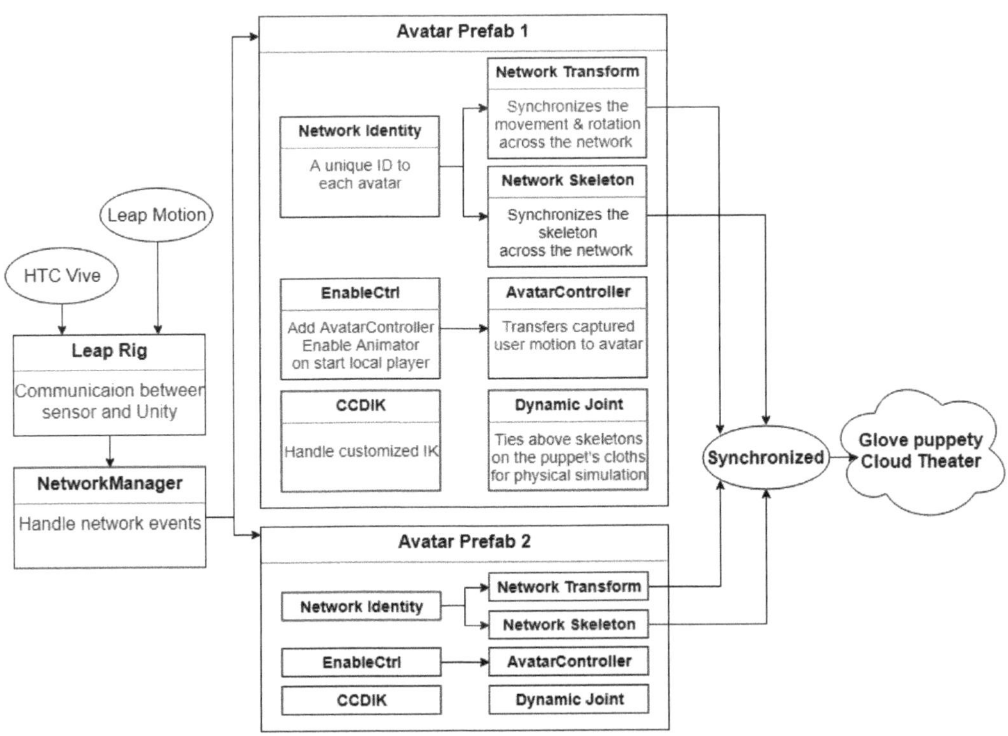

Figure 11. System architecture of VR glove puppetry system.

This paper proposed a multi-user, cloud-based VR system which provides different users with the same virtual scene from their own computers. The system enables users to enter the virtual environment by using a computer. Each user has a unique perspective on the virtual scenes. To communicate with others in the virtual environment, users must consider each other's perspectives and experiences of the scene.

5. User Experience Study

This article presents a study of user experience. In general, the system usability scale [30] is a common reference for usability surveys in case studies for evaluating user experience design. This usability study focuses on conducting user experience research and provides an understanding of the feasible design methodology included in the Chinese puppet cultural preservation VR project. A preliminary study was conducted to evaluate the cloud-based VR prototypes. A group of 30 participants participated in this user experience study, including 14 females and 16 males ranging in age from 21 to 48 years old. This user experience study focuses on measuring the ease of use and usefulness of the proposed VR system.

The test time for this experiment was approximately 30 min to complete the task. First, the experimental conductor explained the wearing method and operation of the equipment with a demonstration. The purpose and process of the experiment was also described in detail, so that the participants could understand the experiment. Each participant wore a head-mounted display, as shown in Figure 12. The experimental process was notified and to be suspended immediately if there was any doubt or discomfort during the experiment. Subjects were invited to provide a questionnaire referring to their experience of using the system after the end of the experiment, which included three parts: basic information, the system usability scale, and glove puppet culture preservation. There were ten questions on the system usability scale, and five questions on the glove puppet culture preservation satisfaction scale. The question was designed using a five-point Likert scale, with 1 point indicating strongly disagree, 3 indicating no opinion, and 5 indicating strongly agree. Participants can also deliver suggestions that they want to provide to this study for the subsequent adjustment of the system.

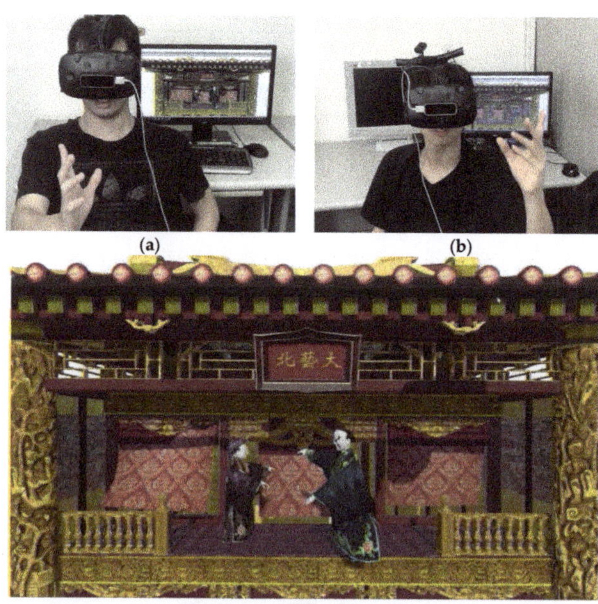

Figure 12. Users experience study sharing the virtual stage of our system. (**a**) User A from place A. (**b**) User B from place B. (**c**) Virtual stage.

The Table 1 questionnaire is the System Usability Scale (SUS) using the Likert five-point scale [30]. SUS produces a single number signifying a measure of the whole usability of the system. SUS scores for individual items are not meaningful. To estimate the SUS score, first sum the score of each item's contribution. The singular in Table 1 is a positive question and the even number is a negative question design. Each item's contribution score will range from 0 to 4. The positive question score is the scale position minus one, and the negative question score is five minus the scale position. The sum of the scores minus 2.5 gives us the overall value of the system usability. The average score is 68 points, which was used as the passing score standard for the system.

Table 1. Questionnaire for the system usability scale [30].

1. I think that I would like to use this system frequently.
2. I found the system unnecessarily complex.
3. I thought the system was easy to use.
4. I think that I would need the support of a technical person to be able to use this system.
5. I found the various functions in this system were well integrated.
6. I thought there was too much inconsistency in this system.
7. I would imagine that most people would learn to use this system very quickly.
8. I found the system very cumbersome to use.
9. I felt very confident using the system.
10. I needed to learn a lot of things before I could get going with this system.

Table 2 shows the results of the cloud-based VR system usability statistics table. The average SUS is 82.08 points, which is higher than the average score of 68 points. In other words, our cloud-based VR system is easy to use and not limited to users of a particular group.

Table 2. Result of the system usability scale.

	Strongly Disagree	Disagree	Neutral	Agree	Strongly Agree	Score
Q1	0	2	3	8	17	3.33
Q2	12	12	4	2	0	3.13
Q3	0	1	4	9	16	3.33
Q4	11	8	7	3	1	2.83
Q5	0	0	5	9	16	3.37
Q6	12	13	4	1	0	3.20
Q7	0	0	3	9	18	3.50
Q8	20	5	4	1	0	3.47
Q9	0	0	5	7	18	3.43
Q10	15	8	6	1	0	3.23

In addition to the survey of system usability, the design of items in Table 3 focused on the following research purposes: (1) whether there is new cultural experience or knowledge learning in the cloud-based VR system; (2) the younger generation's understanding of traditional puppetry and the value of traditional cultural heritage; (3) the experience of two people in a collaborative way to perform a puppet show; (4) the feasibility of the Chinese puppet culture preservation using our cloud-based VR system. Table 4 shows the results of the survey questions on Chinese puppet culture preservation. The scores for all the items were above 4.3. The proposed cloud-based VR system not only provides learning and experience of the operation and performance of puppetry, but also helps to preserve traditional intangible culture. Specially, cloud-based VR brings new cultural experiences or knowledge learning to the post-COVID-19 era.

Table 3. Questionnaire for glove puppet culture preservation.

11. I approached the traditional culture of glove puppetry through the cloud-based VR.
12. I can play glove puppetry with another player through remote collaboration that can facilitate communication and expression to play a show.
13. The content and graphics of the glove puppetry cloud-based VR are same as the traditional glove puppetry.
14. Enhance younger generations' understanding of the value of traditional glove puppetry culture.
15. There are new cultural experiences or knowledge learning with the cloud-based VR in the post COVID-19 era.

Table 4. Result of the Chinese puppet culture preservation.

	Strongly Disagree	Disagree	Neutral	Agree	Strongly Agree	Mean
Q11	0	0	2	9	19	4.57
Q12	0	1	4	8	17	4.37
Q13	0	1	5	7	17	4.33
Q14	0	0	3	9	18	4.50
Q15	0	0	2	10	18	4.53

Furthermore, we asked participants for their comments. On the positive side, users said that "The technology is very interesting, fancy and impressive."; "It gives me a good visual immersive experience."; and "It is great for people new to VR. It has been a great experience". Users also indicated that it "Lacks the tactile feel of puppet physical objects."; that there was a "Problem of force feedback during puppet fighting."; that" Although there are dialogues, it would be better to play sound effects or background music.", and that "The user would like to learn from an expert puppeteer." Occasionally, the quality and stability of the network have an influence on the user experience. Generally, most network VR studies encountered challenges in terms of the server stability.

6. Conclusions and Future Work

It is more challenging to safeguard intangible cultural heritage than tangible cultural heritage because performance art, music, and practices have no distinct form. This paper describes the requirements for interactive experiences, digitization of glove puppetry, and a VR glove puppetry system. Our glove puppetry design was inspired by Chinese glove puppetry history, which includes one stage and many glove puppets. In the process of implementation, a set of operational procedures for comprising modules, data integration, and technology implementation of virtual reality networks are planned for traditional Chinese glove puppetry. The system enables puppetry to transcend time and space and become a dynamic interactive experience. Users without professional training can control glove puppet avatars using their hands. This system involves human–computer and human–human interactions and provides an entertaining experience for users of all ages. The questionnaire design of this study includes the system usability scale and the Chinese puppet culture preservation satisfaction scale. After analysis and statistics, the usability score of this system was 82.08 points, which is higher than the average score of 68 points. This demonstrates that the proposed system has good usability and is easy to use. The satisfaction scores for Chinese puppet culture preservation are high, and our system achieves knowledge sharing and learning regarding cultural inheritance. The system developed in this study contributes substantially to the preservation of glove puppetry.

According to participants' suggestions, some functions need to be enhanced in the future. For action feedback, many subjects were expected to collide in regard to the penetration feedback. Collision detection includes collisions between puppets and between puppets and the background. In addition to the dialogue with each other between users, there should be sound effects or background music. The sound effects showed that the puppet's performance and experience could adhere to reality. However, the users' experience cannot be too long because of hand fatigue. Moreover, the subjects were expected to learn from an expert's puppet skills using our system. In the future, we will provide a performance record system for a grandmaster's puppet.

Looking into the practical applications of the future, the VR glove puppetry system and similar systems can be used to enhance education and increase the accessibility of art forms to the public. VR research also provides the interaction and design for future intergenerational communication for cultural heritage. This system can also be used to promote intangible cultural heritage through interactive VR experiences. It can be used in exhibitions to display intangible cultural heritage and can be promoted based on numbers to promote the popularization of intangible cultural heritage. Intergenerational interaction and learning using our VR system could lead to an exchange between experience and knowledge of traditional culture. The findings of this study can inform the preservation of this form of traditional art. Thus, cloud-based virtual reality has become accessible and reasonable. Moreover, this study's understandings and accomplishments will deliver and relocate this traditional art form, of which we have given a new application.

Author Contributions: Conceptualization, D.-L.W.; methodology, D.-L.W. and Y.-H.W.; software, Y.-H.W.; validation, Y.-H.W. and D.-L.W.; writing—original draft preparation, D.-L.W. and Y.-H.W.; Writing—review & editing, D.-L.W. and Y.-H.W.; supervision, D.-L.W. All authors have read and agreed to the published version of the manuscript.

Funding: The authors would like to thank the Ministry of Science and Technology of the Republic of China, Taiwan, for financially supporting this research under Contract No. MOST 106-2221-E-119-003 and MOST 107-2410-H-119-005.

Informed Consent Statement: All subjects gave their informed consent for inclusion before they participated in the study.

Data Availability Statement: Data available on request.

Conflicts of Interest: The authors declare no conflict of interest.

References

1. UNESCO. Basic Text of the Convention for the Safeguarding of the Intangible. Cultural Heritage (2018 Edition). 2018. Available online: https://ich.unesco.org/en/basic-texts-00503 (accessed on 15 March 2019).
2. Zhao, Z. Digital Protection Method of Intangible Cultural Heritage Based on Augmented Reality Technology. In Proceedings of the ICRIS 2017: International Conference on Robots & Intelligent Systems, Huaian, China, 15–16 October 2017. [CrossRef]
3. Roigé, X.; Arrieta-Urtizberea, I.; Seguí, J. The Sustainability of Intangible Heritage in the COVID-19 Era—Resilience, Reinvention, and Challenges in Spain. *Sustainability* **2021**, *13*, 5796. [CrossRef]
4. Antara, N.; Shuvro, S.E.N. The impact of COVID-19 on museum and the way forward to be resilience. *Uluslararası Müze Egitimi Derg.* **2020**, *2*, 54–61.
5. Markopoulos, E.; Ye, C.; Markopoulos, P.; Luimula, M. Digital museum transformation strategy against the COVID-19 pandemic crisis. In Proceedings of the International Conference on Applied Human Factors and Ergonomics 2021, Virtually, 25–29 July 2021; Springer: Cham, Switzerland, 2021; pp. 225–234.
6. Trunfio, M.; Lucia, M.D.; Campana, S.; Magnelli, A. Innovating the cultural heritage museum service model through virtual reality and augmented reality: The effects on the overall visitor experience and satisfaction. *J. Herit. Tour.* **2020**, *17*, 1–19. [CrossRef]
7. Samaroudi, M.; Echavarria, K.R.; Perry, L. Heritage in lockdown: Digital provision of memory institutions in the UK and US of America during the COVID-19 pandemic. *Mus. Manag. Curatorship* **2020**, *35*, 337–361. [CrossRef]
8. Bekele, M.K.; Pierdicca, E.; Frontoni, E.; Malinverni, S.; Gain, J. A survey of augmented, virtual, and mixed reality for cultural heritage. *J. Comput. Cult. Heritage* **2018**, *11*, 1–36. [CrossRef]
9. Carruth, D.W. Virtual reality for education and workforce training. In Proceedings of the ICETA: IEEE 15th International Conference on Emerging eLearning Technologies and Applications (ICETA), Stary Smokovec, Slovakia, 26–27 October 2017; pp. 1–6. [CrossRef]
10. Liu, D.; Bhagat, K.K.; Gao, Y.; Chang, T.-W.; Huang, R. The Potentials and Trends of Virtual Reality in Education. In *Virtual, Augmented, and Mixed Realities in Education*; Smart Computing and Intelligence; Springer: Berlin/Heidelberg, Germany, 2017; pp. 105–130. [CrossRef]

11. Klauck, R.; Lorenz, S.; Hentschel, C. Collaborative work in VR Systems: A software-independent exchange of avatar data. In Proceedings of the ICCE-Berlin: IEEE 6th International Conference on Consumer Electronics, Berlin, Germany, 5–7 September 2016; pp. 133–136. [CrossRef]
12. UNESCO. Strategy for Training Coming Generations of Fujian Puppetry Practitioners. 2012. Available online: https://ich.unesco.org/en/BSP/strategy-for-training-coming-generations-of-fujian-puppetry-practitioners-00624 (accessed on 15 March 2019).
13. Wu, M.D. Performance Art of Taiwan Glove Puppetry Study. Ph.D. Thesis, National Taiwan Normal University, New Taipei, Taiwan, 2004.
14. Jiang, W.C. A Brief History of Taiwanese Puppet Shows. *J. Chin. Ritual Theatre Folk.* **1990**, *67–68*, 88–126.
15. Jiang, W.C. Development of Taiwanese Puppets after the Recovery. *J. Chin. Ritual Theatre Folk.* **1991**, *71*, 52–69.
16. Wei, Y.W. Reinventing Tradition: Taiwanese Glove Puppetry Art. In *Teaching Asian Art: Content, Context, and Pedagogy*; Chapter 13; Hsieh, K., Cooper, Y.C., Lu, L., Eds.; International Society for Education Through Art: Viseu, Portugal, 2019; pp. 154–196. ISBN 978-989-54830-9-9.
17. Skublewska-Paszkowska, M.; Milosz, M.; Powroznik, P.; Lukasik, E. 3D technologies for intangible cultural heritage preservation—Literature review for selected databases. *Herit. Sci.* **2022**, *10*, 3. [CrossRef] [PubMed]
18. Day, J.M.; Way, D.L.; Chen, K.J.; Lau, W.K.; Hsu, S.C. *A Study on the Virtual Reality of Folk Dance and Print Art—Taking White Crane Dance for Example. Distributed, Ambient and Pervasive Interactions: Understanding Humans*; Lecture Notes in Computer Science (LNCS, Volume 10921); Springer: Cham, Switzerland, 2018; pp. 347–358.
19. Selmanović, E.; Rizvic, S.; Harvey, C.; Boskovic, D.; Hulusic, V.; Chahin, M.; Sljivo, S. Improving accessibility to intangible cultural heritage preservation using virtual reality. *J. Comput. Cult. Herit.* **2020**, *13*, 13. [CrossRef]
20. Huang, W.; Xiang, H.; Li, S. The application of augmented reality and unity 3D in interaction with intangible cultural heritage. *Evolut. Intell.* **2019**. [CrossRef]
21. Han, M.; Yi, Y. Digital Exhibition of Intangible Heritage and the Role of Museums in COVID-19 Era—Focusing on Gwangju Chilseok Gossaum Nori Video Experience Center in South Korea. *Heritage* **2022**, *5*, 2673–2688. [CrossRef]
22. Nitsche, M.; McBride, P.A. Character in Your Hand: Puppetry to Inform Game Controls. In Proceedings of the DiGRA '18: 2018 DiGRA International Conference: The Game is the Message, Turin, Italy, 25–28 July 2018.
23. Lin, C.Y.; Yang, Z.H.; Zhou, H.W.; Yang, T.-N.; Chen, H.-N.; Shih, T.K. Combining Leap Motion with Unity for Virtual Glove Puppets. In Proceedings of the IEEE International Conference on Artificial Intelligence and Virtual Reality, Taichung, Taiwan, 10–12 December 2018; pp. 251–255. [CrossRef]
24. Thomas, J.; Bashyal, R.; Goldstein, S.; Suma, E. MuVR: A Multi-User Virtual Reality platform. In Proceedings of the IEEE 2014 Virtual Reality, Minneapolis, MN, USA, 29 March–2 April 2014; pp. 115–116. [CrossRef]
25. Nguyen, C.; DiVerdi, S.; Hertzmann, A.; Liu, F. CollaVR: Collaborative In-Headset Review for VR Video. In Proceedings of the UIST'17: 30th Annual ACM Symposium on User Interface Software and Technology, Québec City, QC, Canada, 22–25 October 2017; pp. 267–277. [CrossRef]
26. Ibayashi, H.; Sugiura, Y.; Sakamoto, D.; Miyata, N.; Tada, M.; Okuma, T.; Kurata, T.; Mochimaru, M.; Igarashi, T. Dollhouse VR: A multi-view, multi-user collaborative design workspace with VR technology. In Proceedings of the PSA'15: SIGGRAPH Asia, Emerging Technologies, Kobe, Japan, 2–6 November 2015. [CrossRef]
27. Piumsomboon, T.; Lee, Y.; Lee, G.; Billinghurst, M. CoVAR: A collaborative virtual and augmented reality system for remote collaboration. In Proceedings of the SA '17: SIGGRAPH Asia, Emerging Technologies, Bangkok, Thailand, 27–30 November 2017. [CrossRef]
28. Sra, M.; Mottelson, A.; Maes, P. Your Place and Mine: Designing a Shared VR Experience for Remotely Located Users. In Proceedings of the DIS'18: 2018 Designing Interactive Systems Conference, Hong Kong, China, 9–13 June 2018; pp. 85–97. [CrossRef]
29. Seden Puppet Foundation. Available online: http://seden1985.blogspot.com/p/about.html#02 (accessed on 16 March 2019).
30. Brooke, J. SUS: A "Quick and Dirty" Usability Scale. In *Usability Evaluation in Industry*; First Published 1996; CRC Press: London, UK, 2014; pp. 189–194. ISBN 9780429157011.

applied sciences

MDPI

Article

3D Reconstruction of Celadon from a 2D Image: Application to Path Tracing and VR

Seongil Kim and Youngjin Park *

Department of Computer Engineering, Dong-A University, Busan 49315, Republic of Korea; dongaksi7@donga.ac.kr
* Correspondence: yjpark@dau.ac.kr; Tel.: +82-51-200-5824

Abstract: We present a straightforward approach for reconstructing 3D celadon models from a single 2D image. The celadon is a historical example of the surface of revolution. Our approach uses a surface of revolution technique to generate the basic shape of the celadon and then applies texture mapping to create a realistic appearance. The process involves detecting the contour and corners of the celadon image, determining an axis of revolution, generating a profile curve, and finally constructing a 3D celadon model. Additionally, we create models as triangular meshes at multiple resolutions, employing a B-spline curve as the profile curve. It enhances the adaptability of the models for various purposes. We render various scenes using a path tracer to assess the suitability of the generated 3D celadon models and generate a VR celadon museum with the models. Overall, our approach offers a simple and efficient solution for reconstructing a 3D celadon model, generating VR content, and demonstrating extensive applicability across numerous disciplines.

Keywords: 3D reconstruction; surface of revolution; celadon; VR museum; path tracing

Citation: Kim, S.; Park, Y. 3D Reconstruction of Celadon from a 2D Image: Application to Path Tracing and VR. *Appl. Sci.* **2023**, *13*, 6848. https://doi.org/10.3390/app13116848

Academic Editors: Marek Milosz and Jacek Kęsik

Received: 13 May 2023
Revised: 31 May 2023
Accepted: 1 June 2023
Published: 5 June 2023

1. Introduction

Computer graphics technology allows for the effective visualization of digital data and can be used to create digital twins of physical objects for preservation, learning, and research purposes. In particular, 3D digitalized cultural artifacts can provide valuable insights into various fields, such as history, culture, and art. The representative cultural artifact is a celadon [1–3], which is a form of ceramic pottery that originated in medieval East Asia. The celadon helps to understand past cultures, lifestyles, and art styles from when the celadon was originally made. Nevertheless, the celadon is a fragile artifact affected by environmental factors, such as humidity, temperature, and light. These factors can also cause patterns on the celadon to be lost over time, resulting in the decline of its cultural value.

Because of these characteristics of the celadon, researchers have actively focused on 3D reconstruction techniques for its preservation [4–8]. Nowadays, a virtual reality (VR) museum offers immersive experiences that allow users to explore and interact with 3D digitalized artifacts, including celadon pottery [9–11]. Through the VR environment, users can virtually examine celadons from various angles, gaining a deeper understanding of their intricate details and cultural significance [12]. Although sharing and accessing 3D digital models of celadons online is convenient, many people still need help digitizing these valuable cultural artifacts into 3D [11]. To address this challenge, we propose a general guideline that simplifies generating a 3D model and texture from a single 2D input celadon image, making it easier and more accessible. Figure 1 shows the eight 3D models generated by our method from the 2D images, placed on pedestals and encased in glasses as if they were on display in a museum.

Figure 1. The scene $\mathcal{S}_{showcase}$ rendered in 4K resolution with a value of samples per pixel $\sigma = 512$. Our generated 3D celadon models are placed on pedestals and encased in glasses.

From a geometric perspective, the celadon is shaped like a surface of revolution. This unique characteristic simplifies the creation of 3D data compared to other complex 3D objects. Surfaces of revolution can be defined using only two parameters: a profile curve C and an axis of revolution A. Accurately representing the profile curve is the key to successful 3D reconstructions. We first extract a profile polyline using image processing techniques that consider the celadon in the input image as a surface of revolution. The profile polyline can be converted to the B-spline profile curve C by curve fitting [13,14].

B-spline curves are locally defined splines representing various geometry types by adjusting degrees, knots, and control points [15,16]. This property represents the profile curve at different resolutions with fewer data, maintaining the shape of the celadon. Fitting the profile polyline to the B-spline profile curve offers several advantages over other algorithms [17–21] in that the B-spline can offer superior curve representation and accuracy compared to these algorithms. Based on the process, we generate 3D celadon models by rotating the curve around the axis.

The final step is to generate textures that include the colors and patterns of the celadon and apply these to the 3D models. To do so, we first separate a celadon region from a background in the input image. Then, we automatically generate rectangular-shaped texture images using linear interpolation. Applying the generated textures to the celadons will help to analyze and understand them.

There are several ways to represent 3D data, such as triangular mesh, point clouds, voxels, and implicit surfaces. We construct 3D celadon models in the triangular mesh, put the generated textures on them, and render in various scenes using a path tracer [22]. When rendering a scene using a path tracer, selecting a value of samples per pixel (SPP) σ is crucial because it is a trade-off between image quality and rendering speed.

The main contributions of this work can be summarized as follows:

- We propose a general guideline for obtaining a 3D celadon model from one single 2D image without requiring any additional inputs.
- Our method considers the celadon in the input images as a surface of revolution and extracts a profile polyline and an axis of revolution from it.
- Using the fitted B-spline profile curve, we can generate 3D models at various resolutions we want.
- We automatically generate a texture image of the celadon by separating a region of the celadon from a background in the input image and applying linear interpolation.
- We produce various scenes with our 3D celadon models using a path tracer [22] and assess their suitability.
- We also generated a VR celadon museum with the models using Unreal Engine 5, which shows that valuable cultural artifacts can be easily used as VR content and viewed by anyone interested.

2. Related Works

2.1. 2D Image Processing

Suzuki–Abe's algorithm [23] has been widely used for contour detection due to its superior performance compared to an earlier method [24]. They introduced new procedures for border labeling and identifying the parent border of the currently traced border, improving the algorithm's overall accuracy and speed. Moreover, they proposed a method for extracting only an object's the outermost border, which enhances the algorithm's usefulness in various applications, such as object detection, image segmentation, feature extraction, and so on.

Corners are distinct features that can be distinguished from other parts of an image. They are robust to deformations and provide valuable information about the shape and structure of objects. Moravec's method [25] is a classic corner detection algorithm that calculates the intensity variation in small windows shifted in four diagonal directions around a pixel. While the method is suitable for real-time applications due to its simplicity, it is sensitive to noise and may generate false positives when detecting corners in noisy regions.

Harris's method [26] improved the method [25] by looking for regions with significant changes in intensity in multiple directions. The algorithm utilizes the second-moment matrix to compute the corner response, improving noise robustness and offering more reliable corner detection. Shi–Tomasi's method [27] was introduced as an extension of Harris's method [26] that changes the scoring function used to detect corners. It is considered more robust and performs better.

Reducing points in a curve while preserving its shape is crucial in image processing. Various algorithms for polyline simplification have been proposed in the literature [17–21]. For example, the algorithms of Douglas–Pecuker [17] and Visvalingam–Whyatt [21] are threshold-based. Additionally, alternative methods based on B-spline curves have been proposed [13,28,29]. Dierckx [13] proposed a B-spline curve construction method by finding the coefficient of the basis functions that minimize the least-squares error between a given polyline data and the B-spline curve. Hall [28] used a B-spline curve to generate a profile curve for a surface of revolution and generated a 3D model by rotating the B-spline curve around the axis of revolution. Badiu et al. [29] proposed an efficient and accurate technique that generates a B-spline profile curve through photogrammetry and automatically creates the shapes of pottery using CAD.

2.2. 3D Rendering of Surfaces of Revolution (SORs)

Wong et al. [30] proposed a method for reconstructing SORs from a single uncalibrated perspective view by utilizing the characteristics of SORs. Colombo et al. [31] presented a projective geometry technique that utilizes the symmetry properties of SORs for camera self-calibration, 3D reconstruction, and texture extraction from a single uncalibrated image, including SORs.

SORs are common in everyday objects such as bottles, glasses, cans, jars, and pottery. Among these objects, pottery has significant archaeological and cultural value, and the 3D reconstruction of pottery and pottery fragments has been an active research topic. Kampel and Sablatnig [6] proposed an automatic 3D reconstruction method for pottery fragments using point cloud data obtained from 3D scanning. Karasik and Smilansky [7,8] emphasized the usefulness of 3D scanning technology in archeology and proposed a pottery processing automation pipeline for pottery documentation and analysis.

Their approach also involves scanning pottery fragments and restoring the whole pottery into a 3D model using point cloud data. Banterle et al. [4] proposed an automated pipeline for digitizing catalog drawings of pottery types. They segmented the drawings into regions of interest and extracted features from each region. The extracted features are then used to match the drawing with a set of 3D models of pottery types. Dashti et al. [5] presented a virtual pottery system for ceramic artists. The system combines virtual reality and haptic technology to provide a realistic simulation of the pottery-making process. In ray tracing, Kajiya [32] introduced a simplified ray tracing algorithm for SORs that changes the 3D ray–surface intersection problem into a 2D curve–curve intersection problem, which is solved by a strip tree. Baciu et al. [33] suggested hybrid bounding volumes that further developed the strip tree [32] with monotonic interval partitioning.

3. 3D Reconstruction from a 2D Celadon Image

This section comprehensively explains our method for reconstructing a celadon, a representative example of SORs. First, a profile polyline of the celadon can be extracted from a 2D input image. This polyline can be further refined by fitting it to a B-spline curve, allowing for resolution adjustments as necessary. A 3D celadon model can be generated by rotating the curve around the axis of revolution. The following subsections provide detailed explanations of each step in this process.

3.1. Extract a Profile Curve

Extracting a profile polyline from a 2D celadon image is a crucial stage in the process of 3D reconstruction. The contour of the celadon outlines its shape, while its corners help identify distinct features. Furthermore, the axis of revolution is essential in ensuring proper alignment while generating a 3D celadon model. A profile polyline is extracted using the features then fitted to a B-spline curve around the axis to generate the model. Figure 2 illustrates the flow of the proposed method.

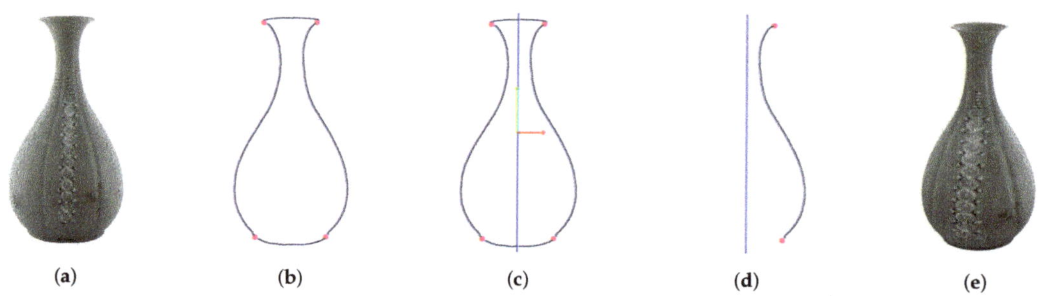

Figure 2. The flow from 2D celadon image to a 3D model: (**a**) input 2D image; (**b**) detected contour and corners; (**c**) the axis of revolution **A**; (**d**) profile curve C with **A**; (**e**) generated 3D celadon model.

Contour detection. Before extracting the contour of the celadon in the input image, we first distinguish it from the background. The binary thresholding method leads us to move a region of interest (ROI) from the whole image to the celadon, so we convert it to a grayscale image and apply the method. After that, we apply Suzuki–Abe's method [23] to the thresholded image to extract an outermost contour from the image and apply Douglas–Peucker's [17] method to approximate the original contour, which enables us to detect referential corners while preserving its shape considerably. The detected contour of the celadon is shown in black in Figure 2b.

Corner detection. A non-flawed celadon always has a rim and a base, with four corners in total, two on the rim and the other two on the base. It is essential to accurately identify the corners of the celadon as they help determine the endpoints of the profile polylines in the contour polyline. However, detecting these corners directly from the original data is challenging. We smooth the contour polyline using a 7×7 Gaussian filter and apply the Shi–Tomasi's [27] method to find its top and bottom corners. These are referential corners used to determine the original contour's corners accurately. The detected corners of the celadon are shown in pink in Figure 2b.

Axis of revolution. The next step is to determine the axis of revolution **A** of the celadon using the four corners. The Principal Component Analysis (PCA) creates two eigenvectors that best describe the original contour. We employ this feature to determine the **A**. Specifically, we select one eigenvector closest to the vertical axis, as it is crucial to align the **A** with the vertical axis. To achieve this, we divide the corners into relatively left and right corners and select any set of them. We can find a direction vector of the **A** using the selected corners and the two eigenvectors. It is important to note that the axis of revolution should pass through the average point of the original contour. If the **A** slightly deviates from the vertical axis, we adjust the **A** and the contour to align with the vertical axis, reducing errors. Figure 2c shows the eigenvectors produced by the PCA in the green and red arrows and the **A** in the blue line.

Profile polylines. The contour, corners, and **A** derived in the previous steps extract the profile polylines. The corners serve as endpoints of them. Selecting one profile polyline between them for generating a 3D celadon model is necessary for the following B-spline curve fitting. Figure 2d shows only the **A** and the celadon's right profile curve. Algorithm 1 summarises extracting a profile polyline from an input celadon image.

Algorithm 1: Extract a profile polyline.

Input : I - an input celadon image
Output: P_p - a profile polyline of the celadon

```
                                        # Contour detection
G_I ← convertToBinaryThreshold(I, threshold)
P_c ← findContour(G_I)
                                        # Corner detection
G_a ← approximateContour(P_c)
G_b ← gaussianFilter(G_a, (7, 7))
F ← findCorners(G_b, P_c)
                                        # Derive an axis of revolution
e_1, e_2, m ← PCA(P_c)
d ← getAxisDirection(e_1, e_2, C)
A ← makeAxisOfRevolution(d, m)
                                        # Select a profile polyline
P_p ← getProfilePolyline(P_c, C, A)
```

3.2. Texture Generation

There are limited patterns and colors present in the input image of the celadon, but it is possible to extract them to create a 2D texture image $T(u, v)$, $0 \leq u, v \leq 1$ to be mapped onto the 3D celadon model. In Section 3.1, we isolated the ROI from the whole image to focus on the celadon. With the ROI, we can generate a rectangular texture image with the same dimensions as the ROI using the scanline method. Non-white color pixels in the ROI are mapped to the texture image while scanning from the celadon's top to bottom. Note that each scanline has a different number of pixels with non-white color pixels. Therefore, we perform linear interpolation on any scanline with a smaller number of pixels than the width of the ROI while mapping.

Figure 3 describes the simplified process of generating a texture image for a simple image, where pixels are shown as circles. Figure 3a shows the input image, while Figure 3b,c highlight a specific scanline with a red border to show the linear interpolation of it. Figure 3d shows the interpolated resulting texture image. Algorithm 2 summarises generating a texture image from an input celadon image.

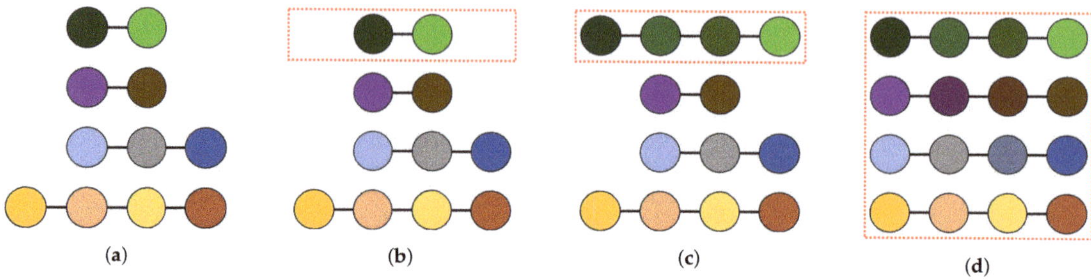

Figure 3. Simplified representation of the texture generation procedure: (**a**) input scanlines; (**b**) selecting one scanline; (**c**) filling the missing data with linear interpolation; (**d**) the final result of texture generation.

Algorithm 2: Generate a texture.

```
Input  : I - a input celadon image
Output: T(u,v) - a texture image of the celadon
ROI ← extractROI(I)
T(u, v) ← makeBlankImage(ROI.dimension)
scanlines ← makeScanlinesOf(ROI)
foreach scanline in scanlines do
    scanline.map(ROI, T(u, v))
    if scanline.width < ROI.width then
        scanline.interpolate(ROI, T(u, v))
    end if
end foreach
```

3.3. Curve Fitting with a B-Spline Curve

The profile polyline obtained from image processing is a discrete polyline on a 2D image. However, such a polyline has resolution limitations in representing a profile curve. To address this issue, it is necessary to transform the profile polyline into a mathematically defined curve, such as a spline. Therefore, we fit the polyline with a third-order B-spline curve by obtaining control points and knot vectors using Dierckx's method [13]. Subsequently, we split the resulting B-spline curve non-uniformly based on the curvature variation to generate a final profile curve C. The profile curve C is then used to generate a corresponding 3D celadon model.

3.4. Construct a Triangular Mesh

The triangular mesh comprises a set of vertices representing the points in $\mathbb{R}^3$ and triangles formed by connecting these vertices with edges. We uniformly sample 360° at a fixed interval and rotate the profile curve C around the axis **A** to generate vertices of the celadon model. Then, we connect the vertices with horizontally adjacent vertices in triangles to create edges and faces.

In addition to generating the 3D model, we should create texture coordinates while connecting the adjacent vertices. The texture coordinates are computed by mapping the angles to the u and the vertical range of the profile curve C to the v. When it comes to mapping for u, we use a linear mapping method, which maps $[0°, 180°]$ to $[1, 0] \in u$ and $[180°, 360°]$ to $[0, 1] \in u$. Finally, we can apply the generated texture image to the 3D celadon model, as shown in Figure 4.

Figure 4. The rendered images of a $\mathcal{P}_0$ model by rotating it 360° in a 60° interval.

4. Experimental Results

Our approach was implemented on a Windows PC with an Intel Core i7-11700 2.5 GHz processor, 32 GB of RAM, and an NVIDIA GeForce RTX 3070 graphics card, using Python 3.11.1. We experimented with eight celadon examples [34] by processing their 2D input images to generate corresponding 3D models. In order to simplify data acquisition, we limited ourselves to using one image per celadon. For rendering the 3D celadon models in various scenes and evaluating their suitability in different environments [35], we employed the Mitsuba 3 renderer [22]. Furthermore, we used Unreal Engine 5 [36] to construct a VR celadon museum exhibiting the models. Figure 5 shows the results of the 3D reconstruction for the celadons, from $\mathcal{P}_0$ to $\mathcal{P}_7$. The first row of Figure 5 shows input 2D celadon images. The second row shows each image's axis of revolution **A** and profile curve C, shown in blue and black, and the third row shows their texture images. The last row shows the generated 3D celadon models for each input image based on their **A**, C, and texture images.

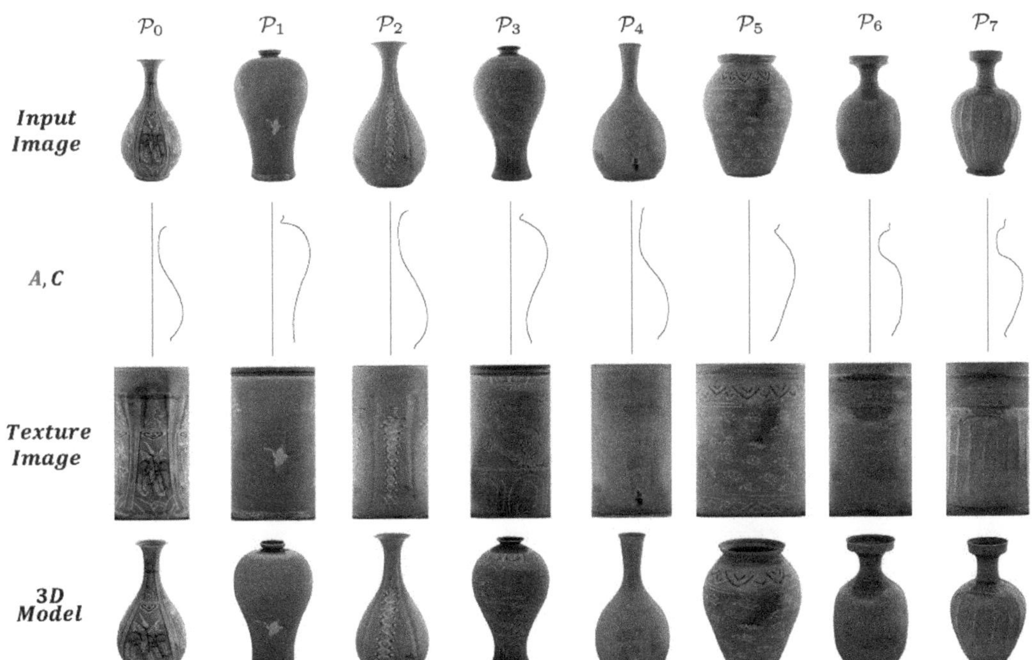

Figure 5. Geometric data generated by our 3D reconstruction algorithm from input image: axis of revolution **A**, profile curve C, generated texture image, and generated 3D model with texture.

Table 1 presents numerical information, such as the computing time for each step of the 3D reconstruction process. Image processing took a similar time for all examples. However, the non-uniformly splitting B-spline curve can impact the resulting 3D model's generation time. As the number of divided domains increases, the number of points and faces in the corresponding 3D model also increases, resulting in slower 3D model generation. The number of divided domains depends on the curvature variations of each profile curve of the celadon, so it has a unique value according to the celadon examples. Meanwhile, we computed the approximation error of the B-spline profile curve C by measuring the one-sided Hausdorff distance to the profile polyline curve with Filip's approximation technique [14]. As shown in Table 1, the B-spline profile curve C approximations are considered to be accurate since their approximation errors are at most 2.33 pixels.

Table 1. Details of 3D reconstruction for each celadon, where the computing times are all measured in milliseconds.

Celadon	Image Processing		B-Spline		3D Model			Approx. Error (Pixels)
	Time (ms)	# Contour Points	# Control Points	# Domains	Time (ms)	# Vertices	# Faces	
$\mathcal{P}_0$	64.06	688	15	261	982.62	94,320	187,920	2.27
$\mathcal{P}_1$	70.12	830	17	298	1076.89	107,640	214,560	2.17
$\mathcal{P}_2$	78.04	835	20	327	1176.80	118,080	235,440	1.38
$\mathcal{P}_3$	68.07	813	18	369	1375.21	133,200	265,680	2.33
$\mathcal{P}_4$	70.03	783	14	382	1469.63	137,880	275,040	1.85
$\mathcal{P}_5$	68.03	699	18	396	1462.05	142,920	285,120	1.45
$\mathcal{P}_6$	69.00	701	24	467	1779.77	168,480	336,240	1.26
$\mathcal{P}_7$	71.01	755	27	554	2082.57	199,800	398,880	1.46

Through several rendering tests, we decided to set the SPP value σ to 64 for low quality and 1024 for high quality and selected HD (1280 × 720), FHD (1920 × 1080), and 4K (3840 × 2160) as the image resolutions to experiment. The scenes were rendered using these two σ values and three image resolutions. Since our GPU hardware does not support σ to 1024 in 4K, we selected 512 for high quality in 4K. The rendering times for each scene at different resolutions and σ values are displayed in Table 2, with each scene represented by a subscript. The results show that the rendering time for each scene is proportional to the image resolution and σ. In particular, the scene $\mathcal{S}_{showcase}$, as shown in Figure 1 with a celadon encased in glasses, is rendered slowly as brighter scenes $\mathcal{S}_{living}$, $\mathcal{S}_{kitchen}$, and $\mathcal{S}_{lounge}$ because path tracing the glass material requires high computational costs. Figure 6a shows the scenes without our 3D reconstructed celadon models, and Figure 6b includes them, with $\sigma = 1024$ and HD resolution. Please note that all scenes were rendered with a GPU and then denoised using an Optix AI-accelerated Denoiser [37].

Table 2. Details of rendering the scenes in various resolutions and σ values, where the rendering times are measured in seconds.

Name	Celadon	Rendering Time (s)					
		HD		FHD		4K	
		$\sigma = 64$	$\sigma = 1024$	$\sigma = 64$	$\sigma = 1024$	$\sigma = 64$	$\sigma = 512$
$\mathcal{S}_{dining}$	$\mathcal{P}_6, \mathcal{P}_7$	2.08	15.97	3.46	35.99	10.08	72.04
$\mathcal{S}_{lounge}$	$\mathcal{P}_1, \mathcal{P}_3$	2.60	39.04	5.84	86.60	22.98	174.00
$\mathcal{S}_{kitchen}$	$\mathcal{P}_0, \ldots, \mathcal{P}_7$	3.87	41.73	7.35	92.75	25.49	187.74
$\mathcal{S}_{showcase}$	$\mathcal{P}_0, \ldots, \mathcal{P}_7$	6.73	102.62	16.18	224.90	57.41	451.41
$\mathcal{S}_{living}$	$\mathcal{P}_0, \ldots, \mathcal{P}_7$	8.90	102.29	15.66	233.02	64.37	469.07

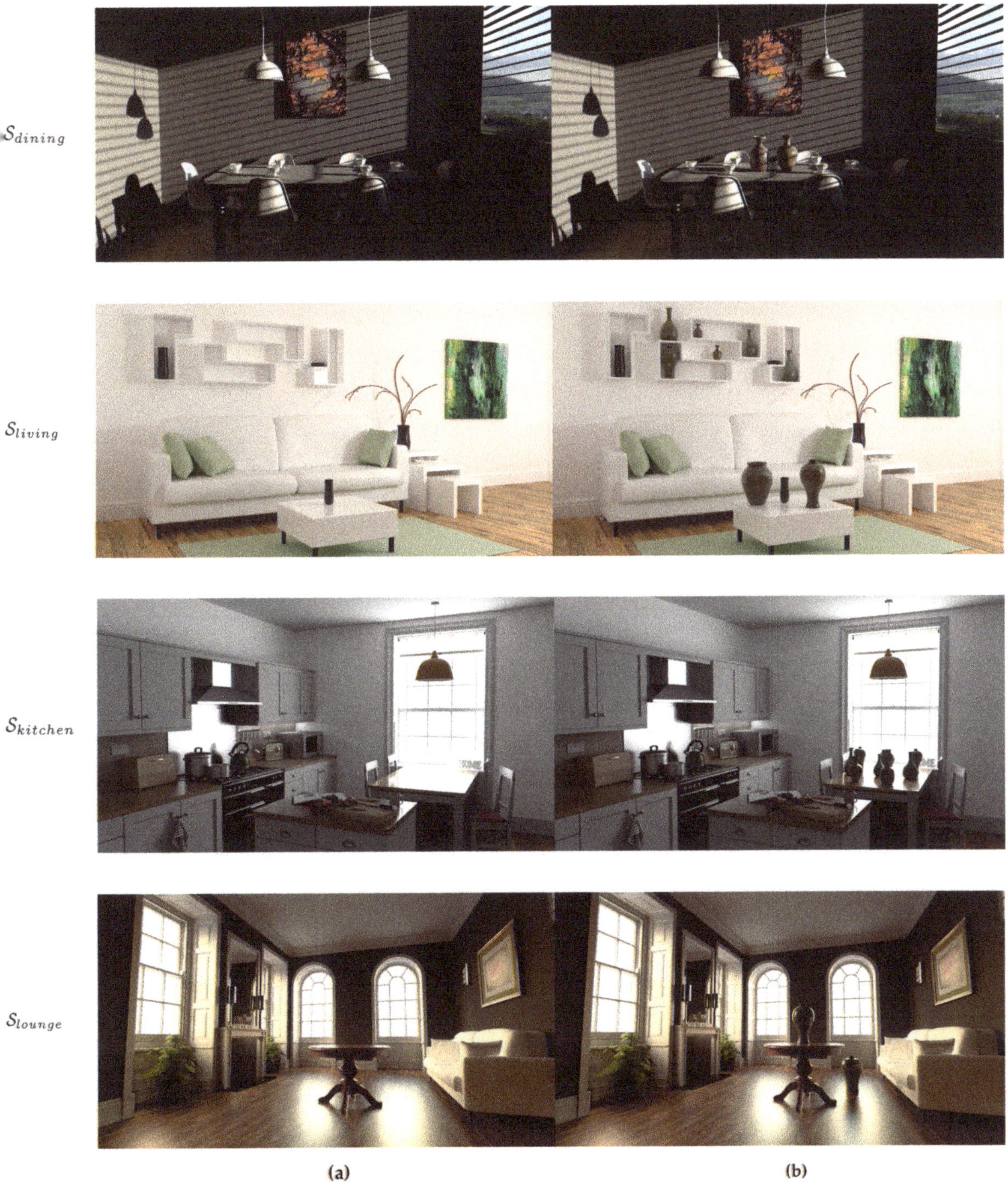

Figure 6. Rendered scenes [35] with a path tracer [22]: (**a**) scenes without our 3D celadon models; (**b**) scenes with the models.

Finally, we generated a VR celadon museum with the celadon models. Figure 7a,b show the layouts of the VR museum in wireframe and an unlit view mode, which are features supported by Unreal Engine 5, and Figure 7c shows a screenshot of the VR museum experience while wearing a VR headset (please refer to the supplementary video for additional details). To enhance the visual aesthetics of the celadon models in the VR environment, we surrounded the models with glass showcases and placed eight directional light sources inside, simulating a real museum environment. Despite being generated from a single 2D image, the models sufficiently represent the geometries and textures of their original celadon, showing that they are suitable for VR content.

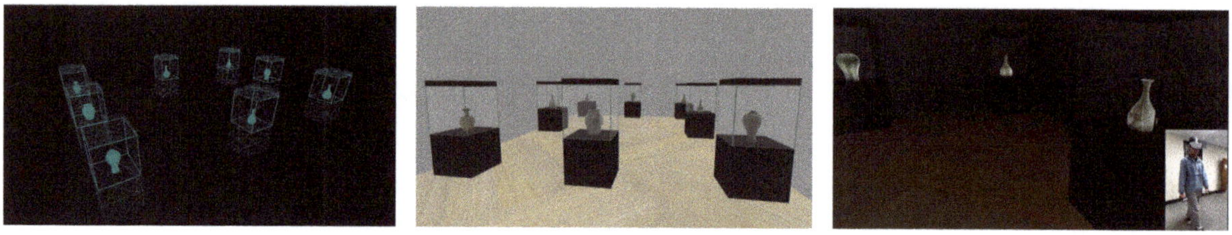

(a) (b) (c)

Figure 7. The VR celadon museum in Unreal Engine 5: (**a**) rendering the museum with wireframe; (**b**) rendering the museum in unlit mode; (**c**) a user wanders around the museum and appreciates 3D celadon models generated by our method.

5. Conclusions

This paper presents a general guideline for generating a 3D celadon model from a single 2D image and further illustrates how to apply the model to various scenes rendered with a path tracer, along with a VR celadon museum. Our approach involves feature extraction from the 2D image and the detection of the profile curve. This curve is approximated by a B-spline curve, which offers a higher curve representation and flexibility compared to other approximation algorithms. This leads to the ability to generate 3D celadon models at any desired resolution. The texture coordinates of the 3D model are also automatically calculated, eliminating the need for any further inputs.

The resulting 3D models can be easily used as VR content, for example, in cultural heritage applications. People can examine the celadon models in the VR celadon museum, facilitating a deeper understanding of the celadon's intricate details and cultural significance. However, the original celadon images were captured via perspective projection with a possibility that celadons can contain specular lightings in the images, causing distortions in the profiles and textures, reducing the reconstruction accuracy. In future work, we plan to calibrate the original images captured via perspective projection and develop an automated method for generating a VR celadon museum.

Supplementary Materials: The following supporting information can be downloaded at: https://doi.org/10.5281/zenodo.7932434 (accessed on 13 May 2023).

Author Contributions: All authors contributed to this work by collaboration. S.K. and Y.P. implemented the proposed method, performed the experiments, and wrote the paper. All authors have read and agreed to the published version of the manuscript.

Funding: This work was supported by the Dong-A University research fund.

Institutional Review Board Statement: Not applicable.

Informed Consent Statement: Not applicable.

Data Availability Statement: The data presented in this study are openly available in Goryeo Celadon Museum, reference number [34].

Acknowledgments: The authors thank Minseok Kim for the technical support of Unreal Engine 5.

Conflicts of Interest: The authors declare no conflicts of interest.

References

1. Carolyn K. Koh Choo. A scientific study of traditional Korean celadons and their modern developments. *Archaeometry* **1995**, *37*, 53–81.
2. Namwon, J. Introduction and Development of Koryŏ Celadon. In *A Companion to Korean Art*; Wiley: Hoboken, NJ, USA, 2020; pp. 133–158.
3. Yan, L.; Liu, M.; Sun, H.; Li, L.; Feng, X. A comparative study of typical early celadon shards from Eastern Zhou and Eastern Han dynasty (China). *J. Archaeol. Sci. Rep.* **2020**, *33*, 102530. [CrossRef]
4. Banterle, F.; Itkin, B.; Dellepiane, M.; Wolf, L.; Callieri, M.; Dershowitz, N.; Scopigno, R. Vasesketch: Automatic 3d representation of pottery from paper catalog drawings. In Proceedings of the 2017 14th IAPR International Conference on Document Analysis and Recognition (ICDAR), Kyoto, Japan, 9–15 November 2017; Volume 1, pp. 683–690.
5. Dashti, S.; Prakash, E.; Navarro-Newball, A.A.; Hussain, F.; Carroll, F. PotteryVR: Virtual reality pottery. *Vis. Comput.* **2022**, *38*, 4035–4055. [CrossRef] [PubMed]
6. Kampel, M.; Sablatnig, R. Profile-based pottery reconstruction. In Proceedings of the 2003 Conference on Computer Vision and Pattern Recognition Workshop, Madison, WI, USA, 16–22 June 2003; Volume 1, p. 4.
7. Karasik, A. A complete, automatic procedure for pottery documentation and analysis. In Proceedings of the 2010 IEEE Computer Society Conference on Computer Vision and Pattern Recognition-Workshops, San Francisco, CA, USA, 13–18 June 2010; pp. 29–34.
8. Karasik, A.; Smilansky, U. 3D scanning technology as a standard archaeological tool for pottery analysis: Practice and theory. *J. Archaeol. Sci.* **2008**, *35*, 1148–1168. [CrossRef]
9. Kabassi, K. Evaluating websites of museums: State of the art. *J. Cult. Herit.* **2017**, *24*, 184–196. [CrossRef]
10. Shehade, M.; Stylianou-Lambert, T. Virtual Reality in Museums: Exploring the Experiences of Museum Professionals. *Appl. Sci.* **2020**, *10*, 4031. [CrossRef]
11. Banfi, F.; Pontisso, M.; Paolillo, F.R.; Roascio, S.; Spallino, C.; Stanga, C. Interactive and Immersive Digital Representation for Virtual Museum: VR and AR for Semantic Enrichment of Museo Nazionale Romano, Antiquarium di Lucrezia Romana and Antiquarium di Villa Dei Quintili. *ISPRS Int. J. Geo-Inf.* **2023**, *12*, 28. [CrossRef]
12. Heeyoung, P.; Cheongtag, K.; Youngjin, P. The Variables of Surface of Revolution and its effects on Human Visual Preference. *J. Korea Comput. Graph. Soc.* **2022**, *28*, 31–40. . [CrossRef]
13. Dierckx, P. Algorithms for Smoothing Data with Periodic and Parametric Splines. *Comput. Graph. Image Process.* **1982**, *20*, 171–184. [CrossRef]
14. Filip, D.; Magedson, R.; Markot, R. Surface algorithms using bounds on derivatives. *Comput. Aided Geom. Des.* **1986**, *3*, 295–311. [CrossRef]
15. Cohen, E.; Riesenfeld, R.F.; Elber, G. *Geometric Modeling with Splines: An Introduction*; CRC Press: Boca Raton, FL, USA, 2001.
16. Farin, G. *Curves and Surfaces for Computer-Aided Geometric Design: A Practical Guide*; Elsevier: Amsterdam, The Netherlands, 2014.
17. Douglas, D.H.; Peucker, T.K. Algorithms for the Reduction of the Number of Points Required to Represent a Digitized Line or Its Caricature. *Cartographica* **1973**, *10*, 112–122. [CrossRef]
18. Lang, T. Rules for the robot draughtsmen. *Geogr. Mag.* **1969**, *42*, 50–51.
19. Opheim, H. Smoothing A Digitzed Curve by Data Reduction Methods. In *Eurographics Conference Proceedings*; Encarnacao, J.L., Ed.; The Eurographics Association: Prague, Czechia, 1981. [CrossRef]
20. Reumann, K.; Witkam, A.P.M. Optimizing curve segmentation in computer graphics. In Proceedings of the International Computing Symposium 1973, Davos, Switzerland, 4–7 September 1973; pp. 467–472.
21. Visvalingam, M.; Whyatt, J.D. Line generalisation by repeated elimination of points. *Cartogr. J.* **1993**, *30*, 46–51. [CrossRef]
22. Jakob, W.; Speierer, S.; Roussel, N.; Nimier-David, M.; Vicini, D.; Zeltner, T.; Nicolet, B.; Crespo, M.; Leroy, V.; Zhang, Z. Mitsuba 3 Renderer. 2022. Available online: https://mitsuba-renderer.org (accessed on 10 January 2023)
23. Suzuki, S.; Abe, K. Topological Structural Analysis of Digitized Binary Images by Border Following. *Comput. Vis. Graph. Image Process.* **1985**, *30*, 32–46. [CrossRef]
24. Moore, D.J.H. An Approach to the Analysis and Extraction of Pattern Features Using Integral Geometry. *IEEE Trans. Syst. Man Cybern.* **1972**, *2*, 97–102. [CrossRef]
25. Moravec, H.P. *Obstacle Avoidance and Navigation in the Real World by a Seeing Robot Rover*; Stanford University: Stanford, CA, USA, 1980.
26. Harris, C.; Stephens, M. A Combined Corner and Edge Detector. In Proceedings of the Alvey Vision Conference, Manchester, UK, 31 August–2 September 1988; pp. 23.1–23.6. [CrossRef]
27. Shi, J.; Tomasi, C. Good Features to Track. In Proceedings of the 1994 Proceedings of IEEE Conference on Computer Vision and Pattern Recognition, Seattle, WA, USA, 21–23 June 1994.
28. Hall, N.S.; Laflin, S. A computer aided design technique for pottery profiles. In *Computer Applications in Archaeology*; Computer Center, University of Birmingham: Birmingham, UK, 1984; pp. 178–188.
29. Badiu, I.; Buna, Z.; Comes, R. Automatic generation of ancient pottery profiles using CAD software. *J. Anc. Hist. Archaeol.* **2015**, *2*.
30. Wong, K.Y.K.; Mendonça, P.R.S.; Cipolla, R. Reconstruction of surfaces of revolution from single uncalibrated views. *Image Vis. Comput.* **2004**, *22*, 829–836. [CrossRef]

31. Colombo, C.; Del Bimbo, A.; Pernici, F. Metric 3D reconstruction and texture acquisition of surfaces of revolution from a single uncalibrated view. *IEEE Trans. Pattern Anal. Mach. Intell.* **2005**, *27*, 99–114. [CrossRef] [PubMed]
32. Kajiya, J.T. New techniques for ray tracing procedurally defined objects. *ACM Siggraph Comput. Graph.* **1983**, *17*, 91–102. [CrossRef]
33. Baciu, G.; Jia, J.; Lam, G. Ray tracing surfaces of revolution: An old problem with a new perspective. In Proceedings of the Computer Graphics International 2001, Hong Kong, China, 3–6 July 2001; pp. 215–222.
34. Goryeo Celadon Museum. Available online: https://www.celadon.go.kr/ (accessed on 20 December 2022).
35. Rendering Resources. Available online: https://benedikt-bitterli.me/resources/ (accessed on 10 January 2023).
36. Unreal Engine 5. Available online: https://www.unrealengine.com/ (accessed on 21 February 2023).
37. Parker, S.G.; Bigler, J.; Dietrich, A.; Friedrich, H.; Hoberock, J.; Luebke, D.; McAllister, D.; McGuire, M.; Morley, K.; Robison, A.; et al. Optix: A general purpose ray tracing engine. *ACM Trans. Graph. (tog)* **2010**, *29*, 1–13. [CrossRef]

applied sciences

MDPI

Article

Optimising Amber Processing Using 3D Scanning: New Perspectives in Cultural Heritage

Sylwester Korga [1,*], Krzysztof Dziedzic [1], Stanisław Skulimowski [1] and Sebastian Gnapowski [2]

1 Department of Computer Science, Faculty of Electrical Engineering and Computer Science, Lublin University of Technology, Nadbystrzycka 36B, 20-618 Lublin, Poland; k.dziedzic@pollub.pl (K.D.); s.skulimowski@pollub.pl (S.S.)
2 Department of Applied Computer Science, Faculty of Mathematics and Technical Informatics, Lublin University of Technology, Nadbystrzycka 38B, 20-618 Lublin, Poland; s.gnapowski@pollub.pl
* Correspondence: s.korga@pollub.pl

Abstract: This article aims to present the practical and functional application of advanced 3D scanning technologies in the process of designing products made of amber. The processing of amber is complex and expensive because mass or serial production is not possible. Moreover, processing is challenging because each product is treated as unique and requires individual handling. The optimisation process is illustrated using the example of amber blocks that were scanned to obtain virtual 3D models. The scope of this work includes the characterisation of three-dimensional jewellery design processes, along with a description of a wide range of methods and functions used in computer-aided jewellery design. The use of computer-aided design software for developing virtual models of decorative products used in goldsmithing is described in this paper. The obtained sets of models underwent engineering analysis using Boolean algebra and data science calculations. In this way, a design methodology was developed regarding the selection of input material as the unprocessed stone for jewellery modelling. The developed method allows for the minimisation of material waste, which is associated with reducing costs in the processing of amber. This article also describes the differences between previous traditional methods and the authors' proprietary method. The proposed methodology can be implemented in jewellery workshops specialising in the processing of materials.

Keywords: amber; 3D scanning; stone processing; computer-aided design; intangible cultural heritage

Citation: Korga, S.; Dziedzic, K.; Skulimowski, S.; Gnapowski, S. Optimising Amber Processing Using 3D Scanning: New Perspectives in Cultural Heritage. *Appl. Sci.* **2023**, *13*, 12973. https://doi.org/10.3390/app132412973

Academic Editor: Andrea Prati

Received: 26 October 2023
Revised: 29 November 2023
Accepted: 30 November 2023
Published: 5 December 2023

1. Introduction

Gemstones and decorative minerals have been used in jewellery, adornments and decorations for centuries, and they hold economic value as well as a significant cultural value. Since ancient times, the processing of stone has been held in high esteem in the hierarchy of craftsmanship, and its underlying knowledge has been shrouded in mystery. Skills such as metalworking have always commanded admiration and respect for this profession [1] and can even be considered as intangible cultural heritage [2]. Therefore, it is crucial to possess knowledge of the properties of the materials being processed, as well as the tools and technologies used in the materials' transformation process.

Materials used in the production of jewellery should possess certain properties such as high hardness, transparency, captivating colour and lustre [3]. This applies to both minerals such as corundum (ruby, sapphire), beryl (emerald, bixbite), nesosilicates (garnet, zircon) as well as organic materials such as amber, jet, or pearl. These features determine the choice of processing methods. In view of the unique properties of each specific piece, the processing of gemstones is complex and expensive due to the absence of mass or serial production methods.

One common challenge in gemstone processing is the susceptibility to mistakes. Typically, each processed stone has an irregular, unique shape, dimensional measurements, and an internal structure. Processing errors are irreversible and often cannot be rectified or repaired [4]. Their characteristics and geometric properties are achieved through various processes, including cleaving, cutting, grinding and polishing [5,6].

The implementation of new technologies that increase efficiency in the manufacturing process and offer products that meet the expectations of a wide range of customers is necessary. The adoption of new technologies in gemstone processing, such as 3D scanning, is also driven by consumers who expect unique and personalised products. Jewellery serves as an example of such products, where the most valued items are unique and not mass-produced [7]. Meeting such demands requires significant effort and financial costs. The labour-intensive process involves designing and constructing models for the creation of moulds and patterns, and making revisions to the design is often impractical. However, with the advancement of computer-aided design and scanning devices, a breakthrough has occurred, enabling the use of tools that were once only within the realm of designers' dreams. Three-dimensional scanning technology allows for the acquisition of virtual models that reflect the shapes of gemstones, allowing work with models in virtual space [8,9]. The use of virtual gemstone and jewellery models reduces costs (financial investments) in producing a particular item by minimising material losses during the grinding process [10].

1.1. Amber Significance and Characteristics

Research on amber, also called "the gold of the north", is conducted by various companies and research institutions specialising in geology, gemmology, archaeology and chemistry. One renowned institution engaged in research on amber is the Institute of Geological Sciences, Polish Academy of Sciences (PAN). This institute performs detailed geological, mineralogical and chemical analyses aimed at understanding the processes of amber formation and its properties. Additionally, research on amber is carried out by private gemmological laboratories that specialise in mineral and gemstone analyses. These laboratories often provide services related to identifying the origin of amber, assessing its authenticity and analysing inclusions inside it, which can provide information about its history. Because amber is also a significant subject of archaeological research, institutions such as archaeological museums and universities conducting archaeological studies often engage in amber analyses, especially if it is found in an archaeological context. It is worth noting that research on amber encompasses diverse scientific fields and requires collaboration among experts from various disciplines to obtain a fuller picture of its origin, properties and history [11].

Amber is used in the jewellery [12], in visual arts [13] or even as a healing agent in folk medicine [14]. Amber was such a desirable commodity, as evidenced by the recorded traces of the existence of an amber trade route between Northern and Southern Europe [15]. Next to hard coal and salt, amber is recognised as a raw material associated with Poland, inextricably linked to the culture and history of this country [16].

Poland is the world's largest supplier of amber. It is often referred to as the global capital of amber, making amber a part of our national heritage. The Amber Route used to be one of the main international trade routes in ancient times. Our scanning technology does not provide direct protection for amber but allows for its optimal utilisation while minimising losses during processing.

Amber is a hard mineral, and more precisely, a fossilised resin from coniferous trees of the Pinus genus and sometimes resinous deciduous trees (from the legume group), as shown in Figure 1. It occurs in many diverse forms and shapes. They depend mainly on the location of the resin exudation, the shape and size of the fissure, the season, sunlight exposure, the condition and age of the tree and the history of the amber itself. The colour of amber is not constant and changes over time from white (rarest, most valued) light yellow to orange and from red to reddish-brown [17–19].

Figure 1. Comparison of amber [Property of the Amber Museum in Gdańsk, Poland].

Sometimes, inclusions (remnants of plants or insects from the Cretaceous and Cenozoic periods) can be found inside the solidified resin, making such specimens the most valuable, Figure 2 [20].

Figure 2. Examples of inclusions in amber [Property of the Amber Museum in Gdańsk, Poland].

1.2. Traditional Methods of Examining Amber Structure

The process of creating jewellery begins with preparatory work, during which the designer analyses the material characteristics of the gemstones. These types of exhibits are often collected in the form of private collections. An example of such exhibits is shown in Figure 3. The exhibits come from the private collection of Dr Elżbieta Miłosz.

Figure 3. Examples of amber necklaces from the private collection of Dr Elżbieta Miłosz.

Previous research methodologies regarding amber have focused on gathering information about its properties, volume analysis, hardness and optical parameters, Figure 4 [21].

Figure 4. A research stand for testing the character of amber—property of the Amber Museum in Gdańsk, Poland.

They provided general descriptions of the studied material and served as a source of information to support the decision-making process regarding its processing. Over time, additional methods and tools were used for general assessment (for example, the Mohs scale for the quick approximation of mineral hardness) and for a detailed evaluation of this property (sclerometers) [22,23] alongside refractometers to determine refractive index [24,25] and an intense assessment of materials (tomographs and equal baseline camera array for voxel grid creation corresponding to density and inclusion distribution [8]). Complete gemstone identification typically requires multiple tests and techniques, such as spectroscopy, density measurement and the observation of microscopic features.

The process of collecting basic data and assessing the quality of the material generates initial ideas and concepts based on the shape of the stone. When selecting a stone for processing, the challenge is to determine its dimensions in a way that minimises production waste. Additionally, it is important to consider whether the chosen amber (stone) can utilise the visual potential of a given design. These challenges in stone selection significantly impact the work's process, technology selection and the final value of the jewellery. So far, this process has involved creating prototypes, which is time-consuming and complex depending on the project and level of detail [9].

The authors of this work propose using modern methods for analysing gemstones in the field of gemmology. These modern methods aim to overcome the limitations of individual techniques and provide a more comprehensive approach to gemstone analysis and identification [26].

2. Methodology and Research

A comprehensive 3D scanning methodology for amber is presented in Figure 5.

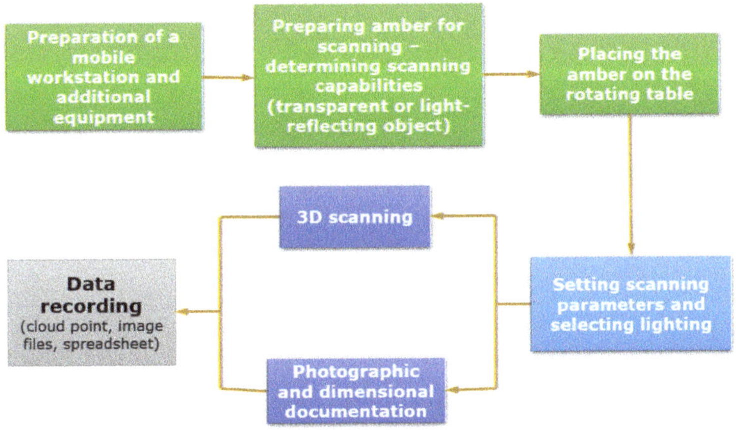

Figure 5. Scanning methodology.

The technology proposed by the authors to support the processing of amber is three-dimensional scanning technology. The scanning setup consists of a handheld scanner using structured light, such as the Artec Spider, a laptop, and a turntable—Figure 6a. The basic parameters of the scanner are as follows:

- Blue light technology scanner.
- Light source: blue LED.
- Three-dimensional resolution, up to 0.1 nm.
- Three-dimensional point accuracy, up to 0.05 nm.
- Data acquisition speed up to 1 million points/s.
- Texture resolution 1.3 MPX (full-colour 3D replicas).
- Working distance 0.2–0.3 m.

Three-dimensional scanning and subsequent data processing require the use of a high-performance laptop. Its basic specifications are as follows: 12-core CPU, 16 GB RAM, 4 GB VRAM and 512 GB SSD. The software used for scanning and data processing was Artec Studio 15.

In the initial scanning phase, it is necessary to select a suitable group of stones that potentially meet the requirements of shape, dimensions and inclusions. Such a group of stones should be both workable and structurally dense enough not to change shape during subsequent operations. Due to the limitations of structured light scanners in scanning transparent and light-reflecting surfaces, the surface of amber was dulled with talc to avoid any additional calculations related to the phenomenon of refraction [9,27].

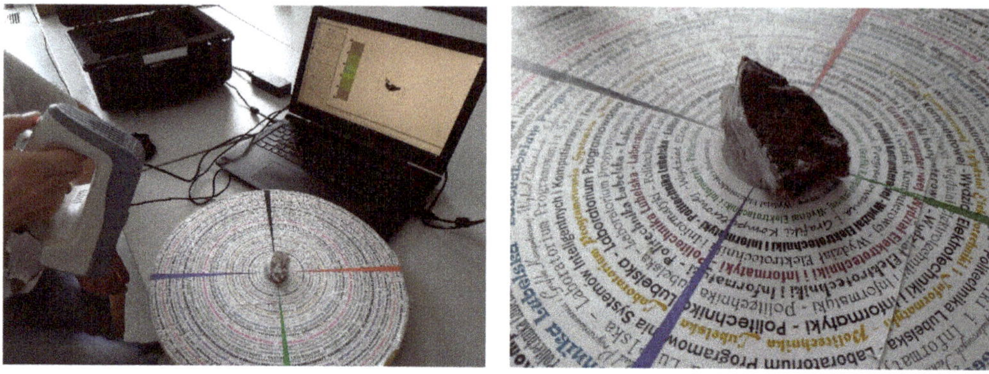

(a) (b)

Figure 6. View of the handheld scanner scanning station: (**a**) general view, (**b**) view of amber on the rotating table.

The models were placed on a turntable, as shown in Figure 6b. The use of a turntable facilitates the complete scanning of the model from all angles. The process of scanning a single component of amber required several to a dozen individual scans, which were then assembled into a complete model during the post-processing stage. The data were collected at a rate of 30 frames per second. The scanning was conducted under natural daylight without the use of additional lamps [28].

The scanning data of object surfaces was saved in the form of a point cloud. During scanning, each scan was subjected to an initial visual assessment. Moreover, the amber was documented through photographic means, and the fundamental dimensions were measured using calipers to verify the precision of the 3D scan. Other properties of amber (Appendix A), such as density distribution, clarity and colour, fraction or weight, were omitted.

2.1. Processing Scanned Models

The data (post-)processing workflow is a multi-stage process, as depicted in Figure 7.

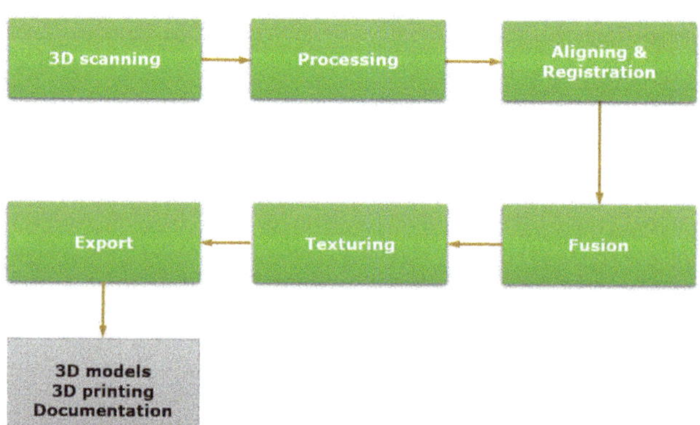

Figure 7. Stages of point cloud processing.

Scanning involved acquiring several to a dozen individual scans that collectively covered the entire surface of the scanned object. The data were imported into Artec Studio 15 software in the form of point clouds with (x, y, z) coordinates and textures.

The first stage of data processing involves aligning each scan to the coordinate system and cleaning it from artefacts, such as scanned fragments of the table, reflections and unwanted elements (Figure 8).

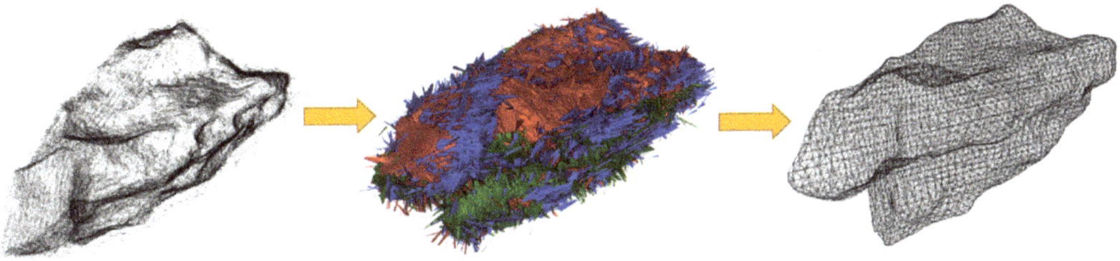

Figure 8. Post-processing steps, from the left: point cloud of an individual scan, alignment of scans, generated mesh model.

In the next stage, Aligning and Registration, involved aligning and merging individual scans to form a unified whole. Automatic alignment options can be used, but the manual selection of common points on the scans is often necessary, requiring at least three pairs of such points. Registration involves bringing the scans into a single coordinate system and optimising their alignment.

The fusion stage involves triangulating the scans after alignment. The Sharp Fusion algorithm was applied to achieve sharp edges in the model. The result was a 3D mesh model of the amber composed of triangles. In the case of missing parts in the mesh, a hole-filling procedure was performed.

For visualisation purposes, the texturing stage involved applying the captured texture to the mesh model. A texture resolution of 2048 × 2048 ppi was used, and brightness and contrast values of the texture display could be adjusted. The final stage was the export of 3D mesh models of the amber.

The OBJ or STL formats were used, which archived the geometry of the models, serving as their digital documentation. In the case of the STL format, it enabled 3D printing.

2.2. Results of the 3D Scanning Process

The 3D scanning and subsequent data processing resulted in digital 3D models of the amber specimens. Table 1 presents the outcomes in the form of mesh models and models with applied texture and their measures. Amber 2 was exceptionally difficult to scan due to the high porosity of the surface, which is why it required a larger number of scans. The scans themselves had to take longer than Amber 1 and 3. As a result, for a constant number of scans 30 fps at the same resolution, an approximately three times denser mesh was produced for Amber 2. The mesh analysis of all models showed the same representation of the surface detail of each object.

The obtained models serve as digital documentation of amber geometry, enabling the planning of amber processing while taking into account the optimal utilisation of the material, such as verifying cross-sections along defined cutting planes.

To present the optimisation process of the processing workflow, the most challenging stone among the available ones was selected—Amber 3. In the vector graphics program Rhinoceros3D, a simplified diamond-shaped model (truncated hexagonal asymmetric bipyramid) was developed for the stone-amber. This model was then copied multiple times, and each copy was scaled down according to a graduated scale. For this purpose, four scaled models were selected. The models of the jewellery were set as solid objects, while the scanned amber model was set as mesh. This allowed for spatial visualisation of the intersecting solids.

Table 1. Volume and surface area of the 3D amber models obtained in Artec Studio.

Object Name	3D Mesh Model	3D Textured Model	Model Triangles	Model Volume [mm^3]	Model Surfaces Area [mm^2]
Amber 1			10,790	13,759	4006
Amber 2	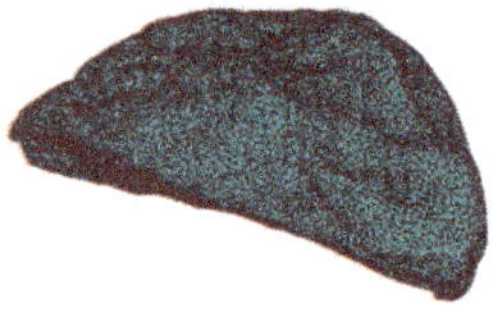		41,448	17,695	4343
Amber 3	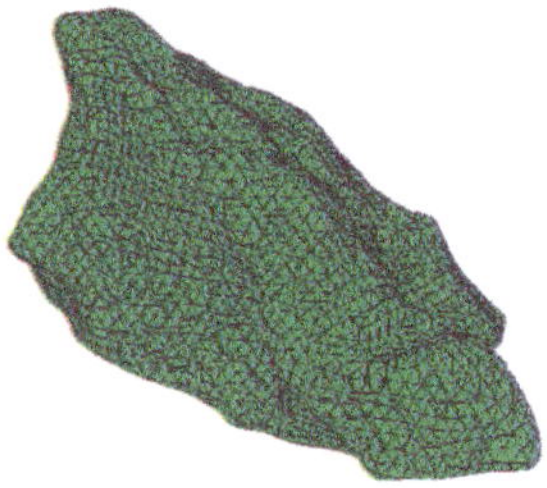	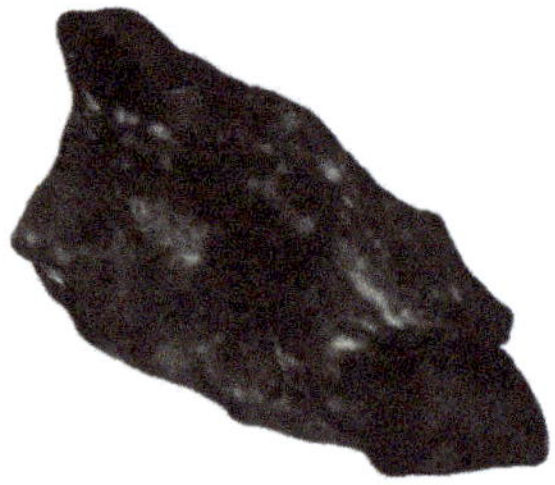	8362	11,495	3142

The mesh of the amber model has an irregular and asymmetric shape, which makes it difficult to overlay the solids. Hence, during this process, four perspectives were employed, namely the top view, front view, right view and perspective view. Working in this arrangement allowed the processed amber model to be placed within the unprocessed amber model. Figures 9–12 illustrate the process of overlaying the spatial models of dodecahedrons and the scanned Amber 3 model.

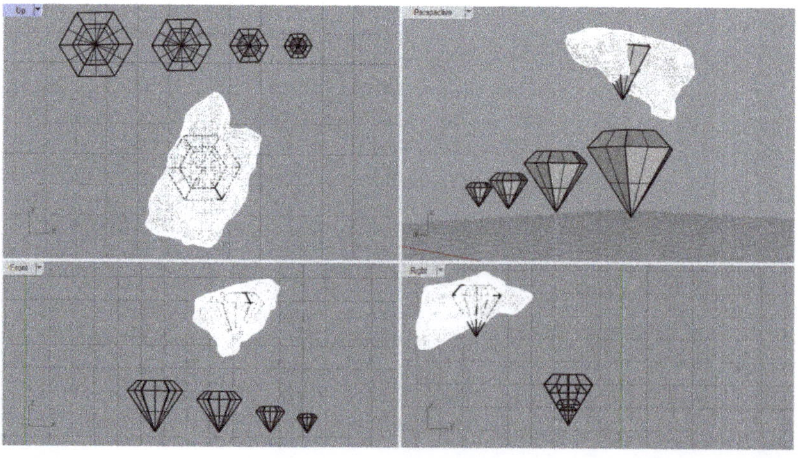

Figure 9. Spatial views in the process of aligning solids (1).

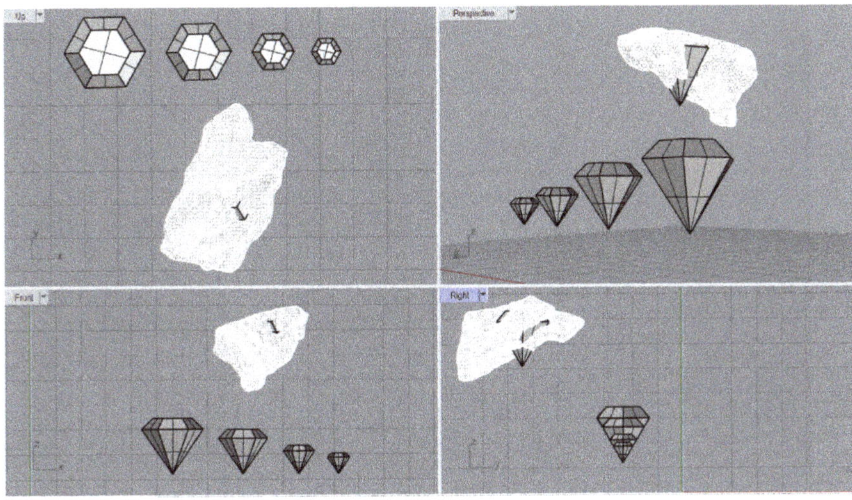

Figure 10. Spatial views in the process of aligning solids (2).

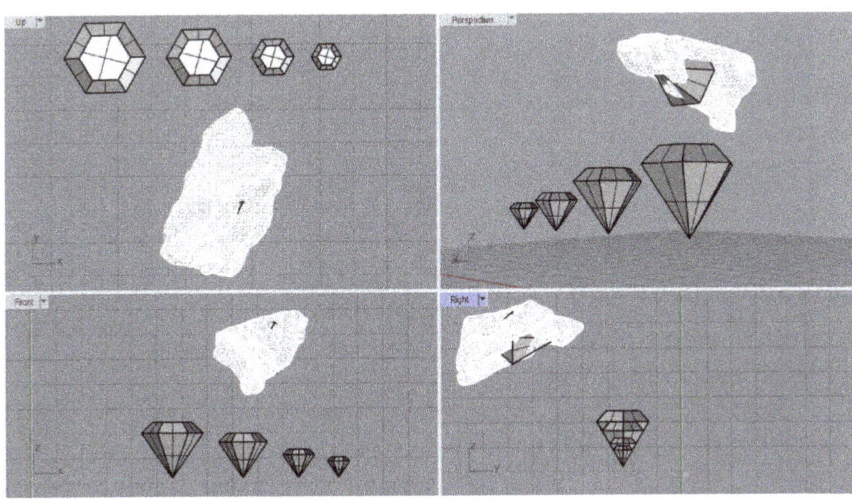

Figure 11. Spatial views in the process of aligning solids (3).

In Figure 12, the optimal position of the diamond model within the mesh of the unprocessed amber is observed. The allowances for grinding were considered during the alignment of solids, which is why the edges of the models do not touch each other directly. Based on the obtained alignments of solids, it is possible to determine how to make cuts in the stone and how to secure it in the holders of saws and grinders. The use of 3D scanning technology and vector graphics provided information for each of the stones about the material losses that could occur during processing. A computational method based on Boolean algebra was used for this purpose. By performing mathematical operations based on volume differences, a comparison of material losses was obtained—Tables 2–4. In each table, Solid No. 6 represented the most favourable alignment model of the diamond model regarding the amber stone.

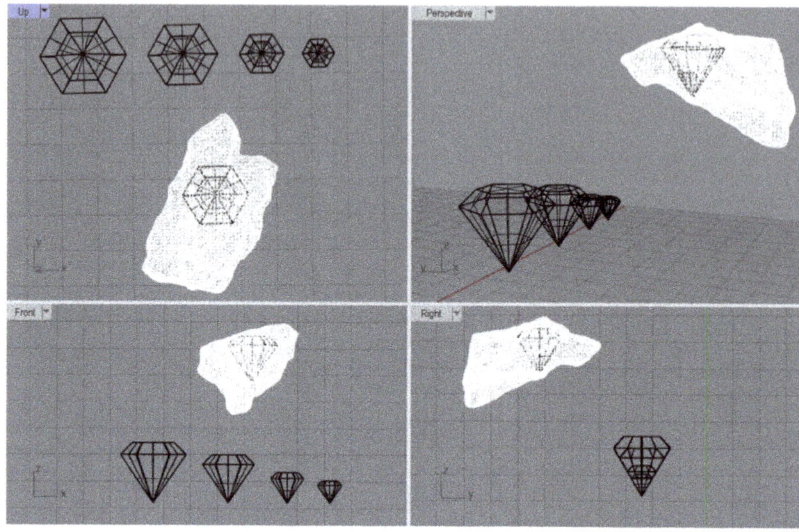

Figure 12. View of the final alignment of solids.

Table 2. Summary of results regarding material losses related to the utilisation of Amber No. 1.

Gem Model			Amber No. 1	
Number of the Solid	**Scale [%]**	**Volume [mm³]**	**Amber Volume [mm³]**	**Material Waste Obtained [mm³]**
Solid 1	100	11,984	13,759	—
Solid 2	80	9587.2		—
Solid 3	60	7190.4		6568.6
Solid 4	40	4793.6		8965.4
Solid 5	20	2396.8		11,362.2
Solid 6	75	8988		4771

Table 3. Summary of results regarding material losses related to the utilisation of Amber No. 2.

Gem Model			Amber No. 2	
Number of the Solid	**Scale [%]**	**Volume [mm³]**	**Amber Volume [mm³]**	**Material Waste Obtained [mm³]**
Solid 1	100	11,049	17,695	—
Solid 2	80	8839.2		—
Solid 3	60	6629.4		11,065.6
Solid 4	40	4420.2		13,274.8
Solid 5	20	2210.1		15,484.9
Solid 6	67	7404		10,291

Prism number 6, included in each table, represents the most favourable matching model of the dodecahedron regarding the amber stone.

This method is implemented at Golden Amber, the world's largest supplier of amber raw materials located near Królewiec/Königsberg. The company is involved in both amber extraction and processing.

Table 4. Summary of results regarding material losses related to the utilisation of Amber No. 3.

	Gem Model		Amber No. 3	
Number of the Solid	**Scale [%]**	**Volume [mm^3]**	**Amber Volume [mm^3]**	**Material Waste Obtained [mm^3]**
Solid 1	100	2050	11,495	—
Solid 2	80	1640		9855
Solid 3	60	1230		10,265
Solid 4	40	820		10,675
Solid 5	20	410		11,085
Solid 6	86	1763		9732

3. Summary and Conclusions

This paper provides information on the utilisation of the 3D modelling process for jewellery design. It compares traditional methods used in gemstone processing and proposes innovative methods utilising 3D scanners and graphic technologies. This study demonstrates that modern jewellery design and stone processing methods significantly contribute to reducing material waste and financial losses. The capabilities of the Rhinoceros 7 software allow for the design of various prototypes with options for saving, editing, scalability and customisation, making it possible to create personalised and unique jewellery designs for each customer. Rhinoceros software enables the creation of diverse jewellery models and the direct export to an STL file format. With this software, any model can be scaled and edited, which is a time-consuming process with traditional techniques. The design capabilities provided by Rhinoceros and Artec Studio allow for the production of customised prototypes in a significantly shorter time tailored to individual customer needs. Compared to traditional jewellery design methods, creating a personalised design is much more cost-effective. The proposed solution may help to strengthen the intangible cultural heritage of amber craftsmanship by relieving craftsmen from at least some hard work so that they can devote the saved time to perfecting the final work stages.

The benefits of using modern technologies in amber processing include the following:

- The rationale behind the incorporation of amber into our research is underpinned by its status as a vital cultural heritage of Poland. Amber serves as a fundamental material for jewellery production by the collaborating companies and enjoys global recognition as a distinct Polish product. It is imperative to note that amber diverges from crystalline structures as it comprises predominantly resin, resulting in polymerisation rather than crystallisation. Within amber, inclusions can be dichotomised into the following two distinct categories: favourable inclusions, encompassing items such as insects, plant matter and water droplets, which appreciably augment the value of amber. Conversely, the second category comprises unfavourable inclusions, including sand, impurities and fractures, leading to a depreciation in the overall value of the amber specimen.
- Improved precision: the advancement of precision is facilitated by modern technologies, such as 3D scanning and modelling, which enable highly precise measurements and calculations, thereby ensuring the precise shaping and cutting of the amber material.
- Waste reduction: the 3D modelling software—Rhinoceros 7 can be utilised to optimise the placement of cutting patterns and designs on the amber, minimising material waste (both suitable for production but less profitable particles (amberoid—pressed amber) and unsuitable dust) and maximising the utilisation of the raw material.
- Time efficiency: the use of modern technologies speeds up the overall processing time. Three-dimensional scanning eliminates the need for manual measurements, while computer-aided design software enables quick adjustments and modifications to the designs [29].

- Enhanced design possibilities: advanced software programs like Rhinoceros provide a wide range of design options and allow for intricate and detailed creations. This enables the production of more complex and unique jewellery pieces using amber.
- Cost savings: Through efficient material usage and reduced processing time, the application of modern technologies can lead to cost savings in the overall production process. Additionally, the ability to create customised designs on-demand minimises inventory costs and reduces the risk of unsold stock. The largest amber processing companies are located in China and Taiwan, and they are also interested in incorporating our method into their workshops. The benefits of this method are self-evident. Reduced waste during processing results in lower material losses, and amber is sold by weight, similar to gold being sold by grams.
- Preservation of original material: the use of non-intrusive 3D scanning techniques aids in preserving the integrity of the original amber material by minimising physical handling and the risk of damage.
- Improved documentation and replication: digital models and records created through 3D scanning and modelling provide comprehensive documentation of the amber pieces. This enables the easy replication of designs and facilitates the production of consistent and identical jewellery pieces.

The next stage of research may be to consider other features, such as the density distribution, clarity or weight, of the tested object in the multi-criteria evaluation process and the financial balance in the process of minimising losses during the decomposition of the facility. Another development may be to increase the level of interaction with scanned models using mixed reality software and technologies such as X-ray [9,30].

Author Contributions: Conceptualisation, S.K. and S.G.; methodology, K.D.; software, S.S.; validation, S.K., K.D. and S.S.; formal analysis, K.D.; investigation, S.S.; resources, S.G.; data curation, S.K.; writing—original draft preparation, S.K.; writing—review and editing, S.S.; visualisation, K.D.; supervision, S.K.; project administration, S.G.; funding acquisition, K.D. and S.K. All authors have read and agreed to the published version of the manuscript.

Funding: This research received no external funding.

Institutional Review Board Statement: Not applicable.

Informed Consent Statement: Not applicable.

Data Availability Statement: Data available on request due to restrictions eg privacy and ethical.

Conflicts of Interest: The authors declare no conflict of interest.

Appendix A

The characteristic features and properties of amber include the following:

- Colour—ranging from light yellow to "amber" brown.
- Transparency—transparent, translucent, opaque.
- Streak—white.
- Density—1.
- Hardness on the Mohs scale—2.0–2.5.
- Cleavage—absent.
- Melting point—287–300 °C.
- Refractive index—1.539–1.542.
- Gravity: 1.05–1.1 g/cm^3

References

1. Erbel, S.; Kuczyński, K. *Cold Working*; PWN: Warszawa, Poland, 1977.
2. Krakow City: List of Intangible Cultural Heritage. Available online: https://niematerialne.nid.pl/niematerialne-dziedzictwo-kulturowe/krajowa-lista-niematerialnego-dziedzictwa-kulturowego/ (accessed on 6 October 2023).
3. Buddhue, J.D. Some new carbon minerals kansasite described. *Mineralogist* **1938**, *6*, 7–8.

4. Głowa, W.; Garbacz-Klempka, A. Research on the Production of Medieval Ornaments. *Sci. J. MHMK* **2010**, *2010*, 97–110.
5. Skrzypek, S.; Przybyłowicz, K. *Metal Engineering and Materials Technology*; PWN: Warszawa, Poland, 2019.
6. Thews, E.R. *Die Chemische Fabrik, 1930—Dechema Werkstoffblätter*; Rabald: Berlin, Germany, 1938; p. 443.
7. Fałkowski, W. *To Rule and Dazzle: Gems and Jewelry in Poland in the 16th and 17th Centuries*; ZKW: Warszawa, Poland, 2019.
8. Kaczmarek, A.L.; Lebiedź, J.; Jaroszewicz, J.; Świeszkowski, W. 3D Scanning of Semitransparent Amber with and without Inclusions. In Proceedings of the WSCG 2021: Full Papers Proceedings, 29. International Conference in Central Europe on Computer Graphics, Visualization and Computer Vision, Pilsen, Czech Republic, 17–20 May 2021; pp. 145–154.
9. Li, X.; Li, Y.; Wang, Y.; Wei, X.; Xin, C. Individualized Inlay Design of Abnormal Amber Based on 3D Scanning Technique. *J. Gems Gemmol.* **2022**, *22*, 37–46.
10. Dzikowska, E. *Jewellery of the World*; Bernardinum: Pelplin, Poland, 2014.
11. Usai, A.; Fiano, F.; Messeni Petruzzelli, A.M.; Paoloni, P.; Farina Briamonte, M.; Orlando, B. Unveiling the impact of the adoption of digital technologies on firms' innovation performance. *J. Bus. Res.* **2021**, *133*, 327–336. [CrossRef]
12. Piskorz-Branekova, E. *Folk Jewelry in Poland*; Sport i Turystyka—Muza: Warszawa, Poland, 2008.
13. Neer, R. Amber, Oil and Fire: Greek Sculpture beyond Bodies. *Art Hist.* **2018**, *41*, 466–491. [CrossRef]
14. Arabas, I.; Popiołek, J.; Pienińska, A. Amber—Medicinal and Magical. *Panacea* **2012**, *2*, 32–33.
15. Kosmowska-Ceranowicz, B. Glowing stone: Amber in Polish deposits and collections. *Przegląd Geol.* **2008**, *56*, 604–610.
16. Kęsik, J.; Miłosz, M. *3D Information Technologies in Cultural Heritage Preservation and Popularisation*; MDPI: Basel, Switzerland, 2023; p. 199.
17. Sadowski, E.M.; Schmidt, A.; Seyfullah, L.; Solorzano-Kraemer, M.; Neumann, C.; Perrichot, V.; Hamann, C.; Milke, C.; Nascimbene, P. Conservation, preparation and imaging of diverse ambers and their inclusions. *Earth-Sci. Rev.* **2021**, *220*, 103653. [CrossRef]
18. Penney, D.; Green, D.I. *Introduction, Preparation, Study & Conservation of Amber Inclusions. Biodiversity of Fossils in Amber from the Major World Deposits*; Siri Scientific Press: Rochdale, UK, 2010; pp. 5–12.
19. Pastorelli, G.; Shashoua, Y.; Richter, J. Surface yellowing and fragmentation as warning signs of depolymerisation in Baltic amber. *Polym. Degrad. Stab.* **2013**, *98*, 2317–2322. [CrossRef]
20. Kosmowska-Ceranowicz, B. The scientific importance of museum collections of amber and other fossil resins. *Pr. Muz. Ziemi* **1990**, *41*, 141–148.
21. Bisulca, C.; Nascimbene, P.C.; Elkin, L.; Grimaldi, D.A. Variation in the deterioration of fossil resins and implications for the conservation of fossils in amber. *Am. Mus. Novit.* **2012**, *3734*, 1–19. [CrossRef]
22. Hardy Allen, R. *The Jewellery Repair Manual (Wydanie II)*; D. Van Nostrand Company Inc.: London, UK, 1967.
23. Jaskulski, A. *Fusion 360 Metodyka Projektowania*; PWN: Warszawa, Poland, 2013.
24. Groover, M.P. *Fundamentals of Modern Manufacturing*, 2nd ed.; Willy&Sons Inc.: New York, NY, USA, 2002.
25. Yoshihara, A.; Maeda, T. Spectroscopic characterization of ambers and amber-like materials. *Vib. Spectrosc.* **2009**, *50*, 250–256. [CrossRef]
26. Gnapowski, S.; Gnapowski, E.; Śniadkowski, M.; Ciekanowski, Z. Using of pulsed plasma discharges and shock wave generator to identify the amber. *Ciencia Tecnica Vitivinicola* **2015**, *6*, 12–22.
27. Montusiewicz, J.; Barszcz, M.; Korga, S. Preparation of 3D Models of Cultural Heritage Objects to Be Recognised by Touch by the Blind—Case Studies. *Appl. Sci.* **2022**, *12*, 11910. [CrossRef]
28. Miłosz, M.; Kęsik, J.; Montusiewicz, J. 3D Scanning and Visualization of Large Monuments of Timurid Architecture in Central Asia—A Methodical Approach. *J. Comput. Cult. Herit.* **2021**, *1*, 1–31. [CrossRef]
29. Dziedzic, K.; Korga, S.; Skulimowski, S. Distance Teaching of 3D Model Post-Processing on the Example of Artefacts from the Silk Road Area. In Proceedings of the INTED 2021, Online, 8–9 March 2021; pp. 3927–3934.
30. Grimaldi, D.; Nguyen, T.; Ketcham, R. Ultra-high-resolution X-ray computed tomography (UHR CT) and the study of fossils in amber. In *Studies on Fossils in Amber, with Particular Reference to the Cretaceous of New Jersey*; Grimaldi, D., Ed.; Backhuys Publishers: Leiden, The Netherlands, 2000; pp. 77–91.

applied sciences

MDPI

Article

Moving beyond the Content: 3D Scanning and Post-Processing Analysis of the Cuneiform Tablets of the Turin Collection

Filippo Diara *, Francesco Giuseppe Barsacchi and Stefano de Martino

Historical Studies Department, University of Turin, Via Sant'Ottavio 20, 10124 Torino, Italy; francescogiuseppe.barsacchi@unito.it (F.G.B.); stefano.demartino@unito.it (S.d.M.)
* Correspondence: filippo.diara@unito.it

Abstract: This work and manuscript focus on how 3D scanning methodologies and post-processing analyses may help us to gain a deeper investigation of cuneiform tablets beyond the written content. The dataset proposed herein is a key part of the archaeological collection preserved in the Musei Reali of Turin in Italy; these archaeological artefacts enclose further important semantic information extractable through detailed 3D documentation and 3D model filtering. In fact, this scanning process is a fundamental tool for better reading of sealing impressions beneath the cuneiform text, as well as for understanding micrometric evidence of the fingerprints of scribes. Most of the seal impressions were made before the writing (like a watermark), and thus, they are not detectable to the naked eye due to cuneiform signs above them as well as the state of preservation. In this regard, 3D scanning and post-processing analysis could help in the analysis of these nearly invisible features impressed on tablets. For this reason, this work is also based on how 3D analyses may support the identification of the unperceived and almost invisible features concealed in clay tablets. Analysis of fingerprints and the depths of the signs can tell us about the worker's strategies and the people beyond the artefacts. Three-dimensional models generated inside the Artec 3D ecosystem via Space Spider scanner and Artec Studio software were further investigated by applying specific filters and shaders. Digital light manipulation can reveal, through the dynamic displacement of light and shadows, particular details that can be deeply analysed with specific post-processing operations: for example, the MSII (multi-scale integral invariant) filter is a powerful tool exploited for revealing hidden and unperceived features such as fingerprints and sealing impressions (stratigraphically below cuneiform signs). Finally, the collected data will be handled twofold: in an open-access repository and through a common data environment (CDE) to aid in the data exchange process for project collaborators and common users.

Keywords: 3D scanning; cuneiform tablets; digital imaging; fingerprints; MSII; sealings

Citation: Diara, F.; Barsacchi, F.G.; de Martino, S. Moving beyond the Content: 3D Scanning and Post-Processing Analysis of the Cuneiform Tablets of the Turin Collection. *Appl. Sci.* **2024**, *14*, 4492. https://doi.org/10.3390/app14114492

Academic Editors: Marek Milosz and Jacek Kęsik

Received: 8 May 2024
Revised: 18 May 2024
Accepted: 23 May 2024
Published: 24 May 2024

1. Introduction

1.1. Cuneiform Tablets Documentation

Clay tablets are a writing medium that first appeared in Mesopotamia in the last centuries of the fourth millennium BCE. Although clay might seem to be an inconvenient material for writing, clay tablets remained in use for more than three thousand years and were adopted by state chanceries, administrative structures, and private people in a wide geographical area that includes Mesopotamia, Iran, Syria, the Levant, Turkey, Greece, and Egypt.

In Western Asia and in Egypt, cuneiform characters were impressed in clay by means of a stylus. The dimension, format, and layout of cuneiform tablets varied according to scribal conventions, time, place, and the purpose of the written text. Cuneiform tablets were produced in both rectangular and square formats; they range in size from as little as 3 cm × 3 cm to as large as 40 cm × 25 cm. All tablets have a convex form, and some of them are shaped like pillows. After forming a tablet from damp clay and smoothing the surface, scribes could write on one or both sides of the tablet as well as on the edges. When the writing process was completed, clay tablets were either dried in the sun or baked [1].

The scientific publication of cuneiform tablets has evolved over time, in step with the available technology. In the nineteenth century, cuneiform tablets were published in the form of hand-drawn copies. Epigraphists copied the tablets either at the sites where they were found or in the museums where they were preserved. Although these hand copies were made by experts, they incorporated a high degree of subjectivity depending on what the individual copyist saw on the tablet—especially when the surface of the tablet was damaged and the intricate cuneiform signs were only partially preserved, as was often the case. Scholars had to interpret the traces visible on the tablet, and sometimes they overinterpreted what they saw.

The development of photography offered the possibility of a more objective representation of the written surfaces of clay tablets. Although photography had been invented in the 1820s, it was not commonly used for commercial and artistic purposes until the second half of the nineteenth century. We are indebted to the photographer Roger Fenton (1819–1869) for the first photographs of cuneiform tablets. Fenton founded the Photographic Society, later called the Royal Photographic Society, in 1853. In the following year, he was appointed as the first official photographer of the British Museum in London and immediately started taking pictures of the tablets in the rich collection [2].

Until recently, hand copies and analogue (film-based) photographs remained the only available visual media for recording and studying cuneiform tablets. Today, with the aid of computers and digital photography, several scientific institutions have created data sets of cuneiform tablets. For example, the Cuneiform Digital Library Initiative (CDLI), founded by Robert Englund at University of California, Los Angeles (UCLA), aims to preserve digital images of cuneiform documents in public museums and private collections available on the internet. About 350,000 cuneiform tablets are currently accessible on the CDLI web site [3].

This rich collection of tablets in the form of digital images and data is a very useful tool for researchers. A major drawback of two-dimensional analogue photography is the difficulty of accurately recording cuneiform signs on the surface of curved tablets, which can be inscribed on all six sides. Digital photography and photogrammetry capture 3D images of tablets, eliminating human error on the part of copyists and allowing accurate paleographic studies.

The CDLI team, in collaboration with the Max Plank Institute, has begun scanning cuneiform tablets with a smart scanner produced by Breuckmann [3,4]. Museums around the world are tackling the digitalization of cuneiform tablets using different techniques and instruments [5–7]. For example, a team at the Vorderasiatisches Museum that is working together with researchers from the University of Münster has started scanning the large collection of cuneiform tablets from Babylon, which is preserved in Berlin. They use a reflectance transformation imaging (RTI) dome that is equipped with 64 LED lamps. The captured 3D images are of high quality, but the instrumentation is bulky and not transportable [8].

1.2. Cuneiform Tablets 3D Scanning and Analysis

This research project Italian Strengthening of ESFRI RI Resilience (ITSERR) [9,10], inscribed into the bigger European programme REligious Studies Infrastructure: tooLs, Innovation, Experts, conNections and Centres in Europe (RESILIENCE) in the framework of European Strategy Forum on Research Infrastructures (ESFRI), focuses on 3D documentation of Near East archaeological artefacts: cuneiform tablets, cylindric seals, bullae, and limestone reliefs. In particular, this work is related to the documentation of an analysis of clay tablets, especially for performing metrological analyses and extracting important metadata (related to tablet production and the humans behind the objects). Post-processing analyses (filters and shader application) will be performed on highly detailed 3D models for the purpose of boosting and investigating nearly-visible details on clay surfaces. Then, these data become fundamental for dedicated open-access repositories for data-sharing: the creation of common data environments (CDEs) for generating Digital Informative Twins will spread the knowledge sphere of these important archaeological artefacts. The schematic workflow is described into Figure 1.

Figure 1. Schematic workflow of the project related to the 3D documentation and analysis of Near East archaeological artefacts.

Furthermore, this project is based on the evaluation of portable 3D scanners for documenting cuneiform tablets, seals, and sealings. Portable scanners can easily be transported to museums or excavation sites to record artefacts.

Our testing began with scanning the cuneiform tablets preserved in Turin at the Musei Reali (Figure 2), and we are grateful to Elisa Panero, Curator of the Archaeological Collection, for her continuous support. This collection of cuneiform tablets, the largest one in Italy, contains more than 650 documents.

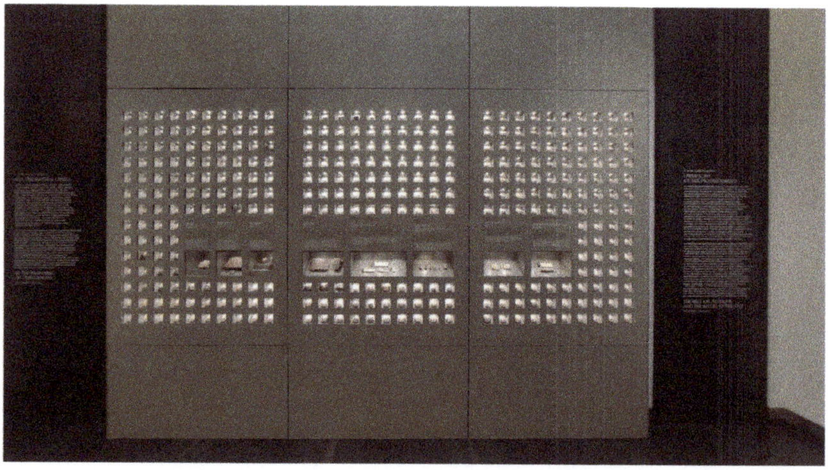

Figure 2. Cuneiform tablet collection, Archaeological Gallery (Musei Reali of Turin—Italy).

All of the tablets have already been edited, transliterated, and translated [11,12]. Analogue photographs of the tablets, along with metadata, are presently available on the web site of the museum, but 3D images capture more detail, enabling a better comprehension of the content of the tablets as well as opening up opportunities for further research, as we will explain later.

The majority of the Turin cuneiform tablets come from the site of Umma in southern Mesopotamia and date back to the last century of the third millennium BCE. They are administrative texts that were issued by officials in the kingdom of the Third Dynasty of Ur.

We have started scanning a group of tablets distinguished by a peculiar feature: all were impressed with seals on one or both sides before they were inscribed [13]. This practice of pre-sealing documents, which was quite common at the time of the Third Dynasty of Ur, had a purpose different from that of sealing a tablet after the scribe had written on it.

A seal impression placed on a tablet that had already been written upon authenticated the content of the document and the identity of its issuer. The practice of sealing a tablet before inscribing it, on the other hand, is in some way comparable to the modern use of letterheads [14].

2. Materials and Methods

In order to perform smart and flexible 3D scans, we adopted the Space Spider scanner from Artec 3D (Luxembourg) [15] (Figure 3). This instrument, a portable handheld scanner, uses a structured-light scanning methodology to record patterns of light deformed by the scanned object. When a beam of structured light strikes the artefact, it is reflected back to the instrument at different angles and dimensions, enabling the scanner to determine the object's morphology [16–18] The Space Spider scanner was chosen not only for its portability, but also for its technical specifications: the 3D point accuracy is 0.05 mm, while the declared 3D resolution of the final output is 0.1 mm. Furthermore, the scanner has a camera sensor onboard (1.3 megapixels) for capturing RGB radiometric values. The scanner's ability to capture these data with accuracy, as well as its flexibility, make the Space Spider one of the best solutions for the 3D documentation of archaeological artefacts, including clay tablets [18].

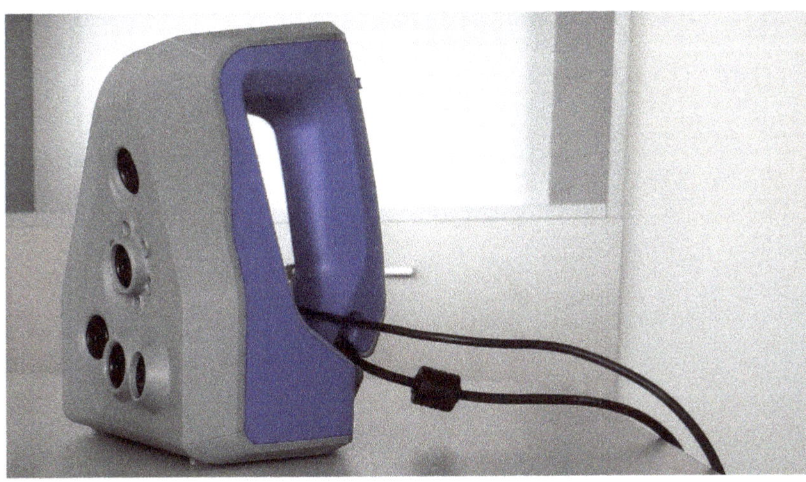

Figure 3. Artec 3D Space Spider structured-light scanner.

2.1. The Cuneiform Sealed Tablet MAT 689

Tablet MAT 689 preserves an administrative document written in the Sumerian language. This tablet, despite its small format (51 mm × 46 mm), was pre-sealed on its entire surface, including the edges (Figure 4). Seal impressions are not always visible to the naked eye and can be seen more clearly in scanned images.

The seal used to make the impressions belonged to the scribe Lu-Haya, who was a member of one of the most distinguished families in the city of Umma. Lu-Haya used several different seals that are documented on multiple tablets from Umma, as reported by Myr 2005 (An updated version of Mayr's PhD dissertation, The Seal Impressions of the Garšana and the Umma Archives, will appear in the series CUSAS (vol. 7) in 2024. Publication is in progress).

Figure 4. Cuneiform sealed tablet MAT 689. Images from the CDLI database. In the figure, verse and obverse were rotated in order to visualize the seal impression correctly.

This tablet was scanned with the Space Spider twice, and a total of 2752 frames were captured. The reconstruction process, via Artec Studio software (version 18.0), returned a 3D model containing approximately 3.6M polygons: fusion processing was set to a 0.08 mm maximum resolution and a 0.16 mm maximum error threshold (Figure 5). Next, the RGB texture layer was computed. By using Artec Studio, the 3D object could be visualized and initially analysed by manually manipulating digital lights. In fact, horizontal and azimuthal light beams can be toggled and manipulated depending on the details to be highlighted. In this way, the individual who analyses the 3D tablet is able to concentrate grazing lights and shadows to enhance micrometric details. Thus, the early analysis of sealings can be performed by changing light directions. The text and figurative scenes on the underlying seal impressions can be better visualized as well.

Figure 5. Three-dimensional model of the cuneiform sealed tablet MAT 689. (**A**) Textured model; (**B**) scan colour model.

2.2. *The Cuneiform Tablet MAT655*

Another administrative Sumerian tablet from Umma, MAT 655, is of particular interest because the 3D scanning operation revealed the fingerprints of the scribe who fashioned the tablet from clay (Figure 6).

Figure 6. Cuneiform tablet MAT 655. Images from the CDLI database.

This tablet was documented by six 3D scans, collecting 3066 frames in total. The final 3D model, with a resolution of 0.1 mm, contained approximately 3.6M polygons. The fusion process was set to a 0.08 mm maximum resolution and a 0.16 mm maximum error threshold. The 3D reconstruction process yielded a high level of detail according to the tablet's dimensions (height, 33 mm; width, 32 mm; depth, 16 mm) and the polygon density ratio. The distinguishing feature of this tablet, the nearly invisible fingerprint, required more transversal scans to capture all the details.

The manually virtual light manipulation using Artec Studio software assisted in the initial detection of surface features; the grooves of the fingerprint were enhanced by switching lights and shadows (Figure 7). Using the measurement tool, the surface area of the fingerprint was computed to be approximately 110 mm^2.

Figure 7. Tablet MAT 655: textured 3D model on the left; scan colour with manipulated light on the right.

2.3. The Envelope Tablet—MAT 740

The Turin collection also preserves some tablets from Mesopotamian sites other than Umma, as well as from other periods. We tested the capabilities of the Spider scanner on an Old Babylonian tablet of unknown provenance that documents a legal act. This tablet, catalogued as MAT 740, is of particular interest because its clay envelope is also preserved (Figure 8).

The clay envelope, which is broken into three pieces, was documented in 3D by 3 separate scans, one for each piece. The reassembled clay envelope measures 87 mm (height) × 55 mm (width), while the inner tablet measures 64 mm (height) × 43 mm (width). A fourth scan session was dedicated to the entire reassembled envelope as an oriented and aligned unique piece to which single scans could be matched. In total, 10,300 frames were collected, and we obtained 3 highly detailed and oriented scans: 2 related to the clay envelope and 1 related to the inner tablet. The entire 3D model of the administrative document contains approximately 23M triangles. Fusion processing was set to a 0.08 mm maximum resolution and a 0.16 mm maximum error threshold (Figure 9).

Figure 8. The envelope tablet MAT 740: (**A**) picture from the archaeological exhibition; (**B**) images of the internal tablet from the CDLI database.

Figure 9. The envelope tablet MAT 740: 3D model of the entire envelope (external tablet) and a particular model of the internal one.

2.4. Post-Processing Filtering

Three-dimensionally documented clay tablets were deeply analysed using post-processing operations based on image computation. In fact, the multi-scale integral invariant filter (MSII) was adopted to extract micro-surface details on polygonal meshes. This filter, implemented inside the open-source software GigaMesh (version 240221) [19–21], enhances the micrometric details and features of 3D models, distinguishing different surface depths and layers.

The obtained 3D models were detailed, including the texture layer, and further analyses were performed during filtering operations using GigaMesh. Although the texture conformed to reality, as is detailed later, filtering and shading operations were carried out without the texture layer because these post-processing computations consider only metric and morphologic features and not RGB colours, which may introduce confusion and prevent an accurate analysis.

The cuneiform tablets presented herein were analysed using an MSII filter with changing parameters depending on surface depths: for example, feature vector element (FVE) parameters and colour ramp visualization were managed in different ways for the purpose of boosting the details of each tablet.

3. Results

Concerning the cuneiform tablet MAT 689, the MSII filter was adopted in two ways: first with 13 feature vector elements (FVEs) and an inverted colour ramp, and then with 5 feature vector elements and a normal colour ramp (Figure 10).

Figure 10. Three-dimensional model of tablet MAT 689: (**a**) scan colour; (**b**) MSII FVEs 13 inverted; (**c**) MSII FVEs 5 normal.

Through this process, the seal impressions in the background could be visualized in different ways by enhancing details on the surface layers. This analysis, via MSII filter, was fundamental for enhancing and revealing unperceived features that lay below the inscribed cuneiform signs. Other filtering processes, such as ambient occlusion and radiance scaling, did not permit such a detailed analysis of the sealings.

Although the sealings on the obverse and reverse were previously visualized and documented by Giovanni Bergamini [12], our post-processing analysis revealed an additional complete seal impression on the edge of the tablet. Bergamini was probably unable to see and analyse this important detail revealed by the MSII filter, and his drawing of the figurative apparatus of the sealing lacks the details on the edge. The analysis proposed herein constitutes an updated and integrative version of the Bergamini drawing (Figure 10).

Post-processing work allows for a clearer reading of the scene engraved on the Lu-Haya seal that was impressed on the tablet MAT 689. A god sits on a two-niched throne, and the god's raised left arm and an astral symbol, undetectable by the naked eye, are now visible. In addition, the post-processed images clearly show that the figure on the left, who is being introduced to the god, is raising both of his arms (Figure 11).

The clay tablet MAT 655 underwent the same post-processing analysis. Filtering the 3D model recovered all micrometric details from surfaces. In fact, the MSII filter was fundamental for boosting the fingerprint grooves and providing a clear vision of all the evidence (Figure 12). This post-processing analysis was designed to set up the MSII 3 feature vector and 13 FVEs. During this analysis, the fingerprint grooves were measured: the average distance between grooves was 0.7 mm, while the depth of a single groove was 0.09 mm.

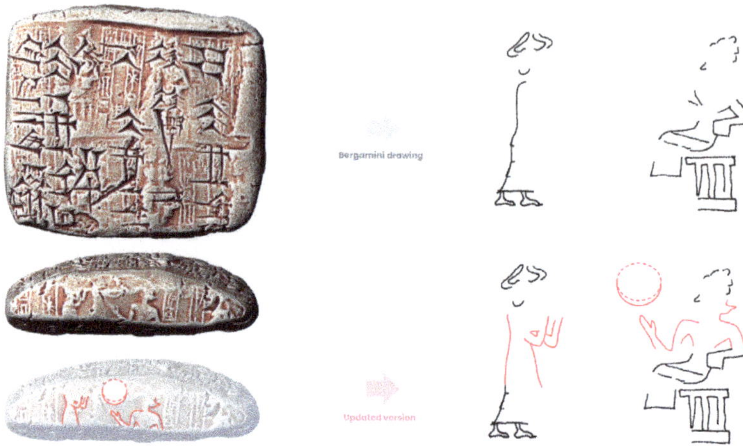

Figure 11. Three-dimensional model of the tablet MAT 689 and sealing analysis: comparison of the original Bergamini drawing and the version obtained via MSII analysis.

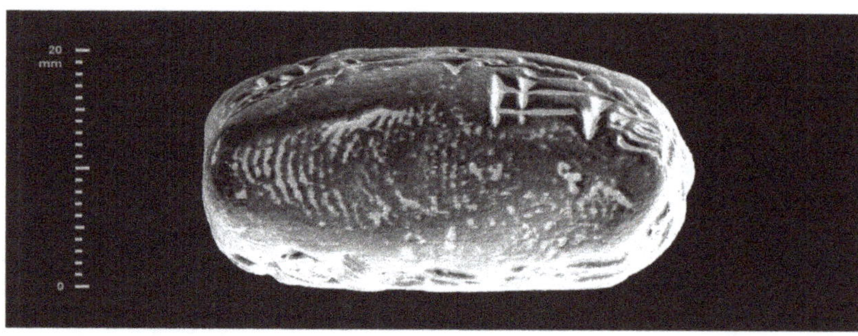

Figure 12. Tablet MAT 655: fingerprint evidence on 3D model acquired via MSII filter (feature vector 3; feature vector elements 13).

The image of the scribe's fingerprint provides a wider perspective on the life of the tablet and establishes a connection with the humans who made these objects. Scanning the entire corpus of the Turin tablets for fingerprints will aid us in distinguishing between the individual scribes who wrote the tablets, even when their names are not recorded in the texts.

As with the tablets analysed previously, the features of the envelope (MAT 740) were first investigated using the light manipulation tool to identify seal impressions and fingerprints, which were enhanced during the MSII analyses. The MSII filter for extracting surface contour features of impressions based on different depth levels was used.

The MSII process, using various analyses of feature vector elements, was fundamental for detecting fingerprint evidence on the edge of the inner tablet—a distinctive feature related to the scribe who made the envelope (Figure 13). The interior of the envelope shows traces in negative form of the signs written on the outside of the envelope, as well as chaotic fingerprint impressions (Figure 14).

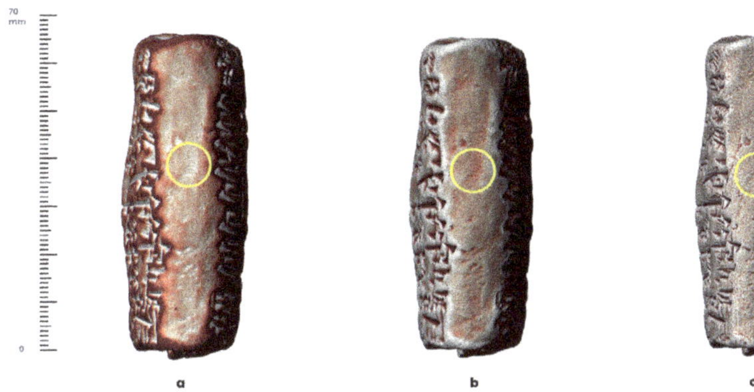

Figure 13. The envelope tablet MAT 740 and the extracted fingerprint (yellow circle): internal tablet. (**a**) MSII filter FVEs 3 normal; (**b**) MSII filter FVEs 3 inverted; (**c**) MSII filter FVEs 13 inverted.

Figure 14. The envelope tablet MAT 740: (**a**) external tablet (side 1) and details of the sealing as well as negative impressions of cuneiform signs; (**b**) external tablet (side 2) and details of the sealing along with chaotic fingerprint impressions.

Metric Evaluation of Space Spider Sensor

Another goal of the WP9 project is related to the metric evaluation of structured-light sensors for the 3D documentation of Near East artefacts. The Space Spider sensor, with a 3D point accuracy of 0.05 mm, is able to reach the deepest angles of cuneiform writing and to perceive the micrometric thickness of seal impressions. In this way, micro details on clay surfaces are properly recorded. For this reason, this structured-light sensor has proven to be one of the best 3D scanners for documenting small- and medium-sized archaeological finds [22–24].

The acquired scans were cleaned by removing frames with deviation errors higher than 0.1 mm (root mean square error—RMSe). Despite the fact that declared and suggested 3D output resolution is 0.1 mm, during the scans, the fusion step can be pushed further to obtain more polygonal density and surface details. However, pushing the declared 0.1 mm resolution too much can sometimes be useless, and the process may create 3D surfaces with artefacts and anomalies.

We tested the fusion process at 0.08 mm with a max threshold error of 0.16 mm, and the three archaeological examples presented herein experienced no anomalies and the surfaces were correctly represented. We also captured more details (Figure 15).

Figure 15. Cuneiform tablet MAT 689. Comparison of 3D models: (**A**) scans fusion set at 0.1 mm max resolution and 0.2 mm max error threshold; (**B**) scans fusion set at 0.08 mm max resolution and 0.16 mm max error threshold.

4. Open-Access Repository and CDE

All relevant information gathered during the scanning and post-processing phases will be collected in a database/open-access repository designed for this purpose.

Most of the data for the database represent two main typologies of objects: cuneiform tablets (and inscribed objects) and sealings. Given the contents of the tablet collection housed in the Musei Reali of Turin, we began compiling the dataset using information on the Sumerian cuneiform tablets from the Ur III period that were recovered from Umma and Drehem [11,12,25]. The dataset meets two essential criteria, which are reflected in very specific sets of queries: basic information and sealing practices.

4.1. Basic Information

The primary focus of the database is to provide a clear and easily accessible overview of all relevant information concerning the physical features of the tablets and sealings that were subjected to 3D scanning. Items are categorized as tablets, seals, or inscribed objects and listed according to their museums or collection numbers. When recording the materials kept in the Musei Reali of Turin, we included both the museum catalogue number and the number assigned to the item in the Cuneiform Digital Library Initiative (CDLI). This facilitates quick comparisons between the information contained in our database and the CDLI repository.

The database field record for each artefact's essential data is related to its place of origin, its size (height, width, and thickness), and its approximate date (if known). Regarding the place of origin, details concerning the country of origin and the main excavation sites are recorded.

Drop-down menus make it easy to retrieve information concerning the material and the state of preservation of each object. We decided to differentiate between three different levels of preservation: complete, fragmentary (some parts of the item are missing), and damaged (the item is complete, but a significant portion of the surface is abraded or damaged). The materials of the tablets and inscribed objects are categorized as clay (in the case of cuneiform tablets) or stone (in the case of seals and other inscribed objects). Further information concerning the materials of artefacts and other features noted during the processing of the scanned images, such as the presence of fingerprints on the surfaces of clay tablets, can be recorded in the remarks field.

4.2. Sealing Practices

The second purpose of the database is to facilitate comparative analyses of particular elements relating to the document or the sealing method. Fields relating to the language of the text and its textual typology respond to this need. The field for recording the content of the text includes a direct link to the transliterations available via the CDLI.

As noted above, 3D scanning and post-processing work reveal details on the surfaces of objects that cannot be detected by the naked eye. Thus, specific fields are dedicated to sealing methods. Two fields are designated for recording the presence of sealings on an artefact and the person who performed the sealing (or rather, the owner of the seal). This allows for very fruitful cross-references with the database of seals and seal impressions hosted by the CDLI and the Database of Neo-Sumerian Texts (BDTNS), which is restricted to the Ur III period [4].

In a free text field, further elements are recorded concerning the number of seal impressions and the orientation and level of the impressions with respect to the inscribed text (i.e., whether the document was pre-sealed or post-sealed), as well as other characteristics that may be identified during a thorough inspection of the sealed surface of the tablet. This field is the place for recording information gained through 3D modelling and post-processing work. For instance, while working on the corpus of Ur III cuneiform documents preserved in the Musei Reali of Turin, we observed that the text of some tablets clearly indicates that the tablet was sealed, even though the tablet itself shows no traces of actual sealings, perhaps because it was enclosed in a clay envelope. The ability to capture invisible surface details in high-definition 3D models could clarify such matters. Finally, a field for recording bibliographical references relevant to the artefacts in the database is provided.

The 3D and semantic data collected from important administrative documents must be securely archived as well as published in order to preserve and disseminate knowledge. Online smart repositories and common data environments (CDEs) could be the proper solution, although they may need to be tailored to specific datasets.

Regarding the project described herein, the tablet dataset deals with different types of information and file formats, ranging from 3D models to 2D images and related metadata. Open-access repositories and CDEs should permit a homogenous view of these data to facilitate complete access and investigation by project collaborators and common users. For this reason, we designed a draft version of an online CDE (Figure 16) for the purpose of analysing cuneiform tablets and their features via a web browser (without any specific hardware limitations).

Figure 16. Draft version of the common data environment for 3D data and metadata on cuneiform tablets.

In the future, this draft version will be implemented by the Italian National Council of Research, thus becoming stable and secure.

From a technical point of view, this environment is essentially composed of a 3D scene coupled with the semantic database apparatus. In this way, 3D models are linked to metadata: descriptions, materials, extracted features, metric information, and the content of the tablet can be easily analysed. Finally, a full-immersion 3D viewer was developed for the purpose of navigating the digital representation of the tablet and performing simplified post-processing analyses.

In the updated version of the repository, relational and semantic queries based on specific ontologies and standards can be performed to investigate historical and extracted metadata and relate them to 3D models.

Designing a smart and open-access environment will facilitate the exchange of semantic data and ease project management and revisions.

5. Discussion

This work focused on the possibility of updating, integrating, and initiating analytic studies on Near East archaeological artefacts. As seen before, clay tablets from Umma enclosed sensitive and extremely important information to be documented and preserved. One of the novelties of this project is related to the possibility of updating seal impression studies, especially to extract the figurative apparatuses as well as to integrate previous studies. Accurate scanning, highly detailed 3D outputs, and MSII analysis could contribute to updating and expanding the investigation and interpretation of sealing procedures.

At the same time, the post-processing analysis of sealings could initiate new semantic connections regarding scribes.

As we have seen, the seal impression of MAT 689 was originally documented by Giovanni Bergamini [12,13], but a deeper and innovative analysis via highly detailed 3D scanning has permitted significant integration of the proposed figurative apparatus.

The extraction of seal impression reveals the methodology behind the marking step: as reported on the tablet MAT 689, sealings were made by both rolling the cylindric seal and impressing a specific area of the textual and figurative apparatus.

Another innovative aspect of this work is related to the possibility of extracting metrically accurate biometric data from clay tablets: After the 3D reconstruction, the MSII filter contributes to extracting and enhancing nearly visible sensitive information from the surface. Fingerprint impressions of scribes can be investigated with micrometric precision, 3D resolution, and an ad hoc post-processing operation. These data may be fundamental for analysing the people behind the objects: In this regard, the age estimation analysis [26–28] and multiple fingerprint matches via fingerprint typology and metric analysis are necessary parameters for identifying the scribes who wrote these documents. However, documenting sub-millimetric details on archaeological finds is always challenging. In the last fifteen years, the 3D documentation method has been designed. Portable structured-light scanners have accelerated the digitization process of risky heritage assets as well as archaeological artefacts permanently preserved in museums.

This research project is also intended to highlight the possibilities of 3D scanning through portable solutions: the Artec Space Spider is revealed as a flexible and smart instrument for quick scanning sessions performed on site (museums). For large projects, a rapid scanning session is currently a much-needed operation. Depending on details and dimensions, the structured-light scanning and early reconstruction of cuneiform tablets requires about 10–15 min. By using other methodologies and instruments, the acquisition and processing time could significantly change: For example, photogrammetric survey is quite swift, but requires time-consuming processing (about 20–30 min, including image masking, image orientation, tie-points extraction, dense cloud, mesh creation, cleaning operations, and texture mapping).

In this regard, the Space Spider sensor, with a 3D point accuracy of 0.05 mm, is an extremely valuable instrument for capturing micrometric details on the surfaces of tablets.

Of course, other commercial scanners like Artec Micro [15], with a 0.01 mm 3D point accuracy and a 0.029 mm 3D resolution, could perform better; however, this solution is not completely portable or flexible.

6. Conclusions

Detailed 3D documentation and precise analysis unlock new possibilities for the investigation of cuneiform tablets beyond the textual content.

A structured-light scanning methodology implemented using a Space Spider scanner facilitated the accessibility and documentation of a tablet collection. Although there are other detailed documentation possibilities (e.g., photogrammetry and RTI Dome), 3D scanning provides quick and highly detailed 3D documentation that is useful for offering different outputs as well as conducting multiple analyses.

As we have seen, the analysis of the tablet MAT 689 highlighted the importance of post-processing imaging for updating existing studies on seal impressions. The three-dimensional modelling of archaeological artefacts should not be an end in itself, but a means for extracting specific data and performing further analyses.

The MSII filtering process shows enormous potential for detecting micro-evidence pertaining to tablet production. Notably, fingerprint analysis can generate broader studies on the people behind the archaeological artefacts by establishing new smart connections between tablets coming from the same site. The entire Turin collection is currently the subject of such research, and in the future, these methods can also be implemented in museums worldwide to generate comparative analyses of particular elements related to fingerprint impressions or sealing methods.

This initial analysis on different clay tablet typologies unlocks new possibilities for future works (some of these are already in progress). Based on these possibilities, a future ongoing work is related to the fingerprint analysis of a cuneiform tablet coming form Puzrish-Dagan (Drehem, Iraq). As mentioned previously, the high 3D point accuracy of actual structured-light sensors, in addition to micrometric analysis via MSII or virtual RTI, provides a wider outlook on fingerprint impressions on clay tablets. Accurate metric investigations can result in more precise studies on age estimation, gender analysis, scribe workshops, and object production.

At the same time, another future ongoing project will concentrate on the analysis of the state of damage and preservation of the limestone relief of the Assyrian King Sargon II, a masterpiece preserved at the Musei Reali of Turin. Through Space Spider scanning, MSII analysis, and virtual RTI, the damaged and erased decoration (common practice in the 19th century) of the Assyrian King will be deeply investigated in order for us to understand the original figurative apparatus of robes.

Finally, the creation of a smart open-access repository and CDE will open up new research possibilities by dismantling data accessibility barriers, allowing full access to and exchange of metadata, 3D models, and extracted semantic data.

Author Contributions: Conceptualization, F.D., F.G.B. and S.d.M.; methodology, F.D.; software, F.D.; validation, F.D., F.D., F.G.B. and S.d.M.; investigation, F.D., F.G.B. and S.d.M.; resources, S.d.M.; data curation, F.D., F.G.B. and S.d.M.; writing—original draft preparation, F.D., F.G.B. and S.d.M.; writing—review and editing, F.D., F.G.B. and S.d.M.; visualization, F.D., F.G.B. and S.d.M.; supervision, F.D., F.G.B. and S.d.M.; project administration, S.d.M.; funding acquisition, S.d.M. All authors have read and agreed to the published version of the manuscript.

Funding: This work was supported by the PNRR (National Recovery and Resilience Plan) project Italian Strengthening of ESFRI RI Resilience (ITSERR) founded by the European Union—NextGenerationEU (CUP:B53C22001770006).

Institutional Review Board Statement: Not applicable.

Informed Consent Statement: Not applicable.

Data Availability Statement: Data are contained within the article.

Acknowledgments: The authors would like to acknowledge Elisa Panero (curator of the Archaeological Collection) and Cristina Meli (Restorer) of the Musei Reali of Turin for their constant support.

Conflicts of Interest: The authors declare no conflicts of interest.

References

1. Balke, T.E.; Tsouparopoulou, C. *Materiality of Writing in Early Mesopotamia*; De Gruyter: Berlin, Germany; Boston, MA, USA, 2016. [CrossRef]
2. Welch, J. Aristocratizing the Community: Roger Fenton and British Photography in the 1850's. Ph.D. Thesis, Princeton University, Princeton, NJ, USA, 2004.
3. CDLI Website. Available online: https://cdli.mpiwg-berlin.mpg.de/about (accessed on 8 May 2024).
4. Tsouparopoulou, C. Progress Report: An Online Database for the Documentation of Seals, Sealings, and Seal Impressions in the Ancient Near East. *Stud. Orient. Electron.* **2014**, *2*, 37–68.
5. Hameeuw, H.; Geert, W. New Visualization Techniques for Cuneiform Texts and Sealings. *Akkadica* **2011**, *132*, 163–178.
6. Kotoula, E.; Akoglu, K.G.; Frahm, E.; Simon, S. QR Coded 3D Prints of Cuneiform Tablets. *Int. J. Art Cult. Des. Technol. (IJACDT)* **2017**, *6*, 1–11. [CrossRef]
7. Antinozzi, S.; Fiorillo, F.; Surdi, M. Cuneiform Tablets Micro-Surveying in an Optimized Photogrammetric Configuration. *Heritage* **2022**, *5*, 3133–3164. [CrossRef]
8. Kleber, K. Die Edition und Erforschung von Neubabylonischen Keilschrifttafeln Aus Babylon. *Alter Orient Aktuell* **2022**, *19*, 4–9.
9. ESFRI RI Resilience Project. Available online: www.resilience-ri.eu (accessed on 8 May 2024).
10. ITSERR—Resilience Associated Project. Available online: www.itserr.it (accessed on 8 May 2024).
11. Archi, A.; Pomponio, F. *Testi Cuneiformi Neo-Sumerici da Drehem, Catalogo del Museo Egizio di Torino. Serie Seconda: Collezioni*; Ministero per i Beni Culturali e Ambientali. Soprintendenza al Museo delle Antichità Egizie: Torino, Italy, 1995; Volume 7.
12. Archi, A.; Bergamini, G.; Pomponio, F. *Testi Cuneiformi Neo-Sumerici da Umma, Parte I: I Testi. Parte II: Le Impronte di Sigillo. Catalogo del Museo Egizio di Torino. Seria Seconda: Collezioni*; Ministero per i Beni Culturali e Ambientali. Soprintendenza al Museo delle Antichità Egizie: Torino, Italy, 1995; Volume 8.
13. Bergamini, G. Gli scribi di Umma. Prassi di validazione del documento e certificazione d'autorità in età neo-sumerica. In *L'ufficio e il Documento. I Luoghi, i Modi, Gli Strumenti Dell'amministrazione in Egitto e Nel Vicino Oriente Antico*; Mora, C., Piacentini, P., Eds.; Quaderni di Acme: Milano, Italy, 2006; pp. 209–220.
14. Laurito, R.; Mezzasalma, A.; Verderame, L. Texts and Labels: A Case Study from Neo-Sumerian Umma. In *Proceedings of the 51st Rencontre Assyriologique Internationale, Chicago, IL, USA, 18–22 July 2005*; Biggs, R.D., Myers, J., Roth, M., Eds.; The Oriental Institute of the University of Chicago Studies in Ancient Oriental Civilizations 62: Chicago, IL, USA, 2005; pp. 99–110.
15. Artec3D Website. Available online: https://www.artec3d.com (accessed on 8 May 2024).
16. Georgopoulos, A.; Ioannidis, C.; Valanis, A. Assessing the performance of a structured light scanner. *Int. Arch. Photogramm. Remote Sens. Spatial Inf. Sci.* **2010**, *38*, 5.
17. Mugnai, F.; Tucci, G.; Da Re, A. Digital image correlation in assessing structured-light 3D scanner's gantry stability: Performing David's (Michelangelo) high-accuracy 3D survey. *Int. Arch. Photogramm. Remote Sens. Spat. Inf. Sci.* **2021**, *46*, 463–469. [CrossRef]
18. Diara, F. Structured-Light Scanning and Metrological Analysis for Archaeology: Quality Assessment of Artec 3D Solutions for Cuneiform Tablets. *Heritage* **2023**, *6*, 6016–6034. [CrossRef]
19. Gigamesh Software. Available online: https://gigamesh.eu/ (accessed on 8 May 2024).
20. Mara, S.H.; Krömker, S.; Breuckmann, J.-B. GigaMesh & Gilgamesh: 3D Multiscale Integral Invariant Cuneiform Character Extraction. In Proceedings of the VAST: International Symposium on Virtual Reality, Archaeology and Intelligent Cultural Heritage, The Eurographics Association, Paris, France, 21–24 September 2010. [CrossRef]
21. Homburg, T.; Zwick, R.; Mara, H.; Bruhn, K.-C. Annotated 3D-Models of Cuneiform Tablets. *J. Open Archaeol. Data* **2022**, *10*, 1–8. [CrossRef]
22. Fisseler, D.; Müller, G.G.W.; Weichert, F. Web-Based Scientific Exploration and Analysis of 3D Scanned Cuneiform Datasets for Collaborative Research. *Informatics* **2017**, *4*, 44. [CrossRef]
23. Göldner, D.; Karakostis, F.A.; Falcucci, A. Practical and technical aspects for the 3D scanning of lithic artefacts using micro-computed tomography techniques and laser light scanners for subsequent geometric morphometric analysis. Introducing the StyroStone protocol. *PLoS ONE* **2022**, *17*, e0267163. [CrossRef] [PubMed]
24. Innerhofer, F.; Reuter, T.; Coburger, C. More Than Just Documenting the Past: 15 Years of 3D Scanning at the Archaeological Heritage Office of Saxony. In *The 3 Dimensions of Digitalised Archaeology*; Hostettler, M., Buhlke, A., Drummer, C., Emmenegger, L., Reich, J., Stäheli, C., Eds.; Springer: Cham, Switzerland, 2024. [CrossRef]
25. Mayr, R.H. Seal Impressions on Tablets from Umma. Ph.D. Thesis, University of Leiden, Leiden, The Netherlands, 1997. Available online: https://www.academia.edu/45128929/Seal_Impressions_on_Tablets_from_Umma (accessed on 8 May 2024).
26. Arslan, A. Studying Fingerprints in Archaeology: Potentials and Limitations of Paleodermatoglyphics as an Archaeometric Method. *Turk. J. Archaeol. Sci.* **2023**, 1–16.

27. Kamp, K.A.; Timmerman, N.; Lind, G.; Graybill, J.; Natowsky, I. Discovering Childhood: Using Fingerprints to Find Children in the Archaeological Record. *Am. Antiq.* **1999**, *64*, 309–315. [CrossRef]
28. Králík, M.; Novotný, V. Epidermal Ridge Breadth: An Indicator of Age and Sex in Paleodermatoglyphics. *Var. Evol.* **2003**, *11*, 5–30.

MDPI AG
Grosspeteranlage 5
4052 Basel
Switzerland
Tel.: +41 61 683 77 34

Applied Sciences Editorial Office
E-mail: applsci@mdpi.com
www.mdpi.com/journal/applsci

www.ingramcontent.com/pod-product-compliance
Lightning Source LLC
Chambersburg PA
CBHW061600210926
51850CB00009B/2160
* 9 7 8 3 7 2 5 8 1 6 1 3 2 *